Sugar Isn't Always Sweet

Sugar Isn't Always Sweet

Maura (Jinny) Zack &

Wilbur D. Currier, M.D.

UPLIFT BOOKS

Brea, California

The names of some persons mentioned in this book have been changed to protect their privacy.

Sugar Isn't Always Sweet

Copyright © 1983 by Maura (Jinny) Zack and Wilbur D. Currier, M.D.

Published by Uplift Books, 428 So. Brea Blvd., Suite B, Brea, CA 92621

Library of Congress Cataloging in Publication Data

Zack, Maura.
    Sugar isn't always sweet.

    Bibliography: p.
    1. Hypoglycemia. 2. Hypoglycemia – Diet therapy.
I. Currier, Wilbur D.   II. Title. [DNLM: 1. Hypoglycemia. WK 880 Z16s]
RC662.2.Z33   1983   616.4'66   83-10354
ISBN 0-88005-002-0

All rights reserved. This book may not be reproduced in any form without violation of copyright.

Printed in the United States of America.

To
Jerry, Julie and Jim
who suffered through it all
and to
Jay who never understood my condition

Acknowledgments

To John and Dawn Beck who provided a home-away-from-home on my many trips to do research at Stanford in northern California. To Faith and John Wightman, Terry and Frank Barrera, and Rita and Larry Keeshan who opened their homes to me when I had to be in southern California for research.

To the many reference librarians who helped with research, especially Leslie Navari in Pacific Grove and Emiko Samard and Steve Clancy at the University of California, Irvine. To Dr. Ruth Ray of Carmel, California, whose answers to my countless questions provided me with much helpful information about hypoglycemia. To Dr. Osman Hull of Monterey, California, for believing in what I was doing, and for his promise to help with RIME. And to all those physicians who reviewed the manuscript for medical accuracy.

To Karen Lee for her many hours and skillful work on the final editing. To Don Tanner for always believing that I had a marketable book, and for his editorial support and professional guidance. To Dr. Wilbur Currier without whose encouragement and collaboration this book may have never been written.

Most of all, to God, for turning my suffering into compassion and the desire for more knowledge.

Foreword

Hypoglycemia, low blood sugar, is a condition which has been estimated to affect as many as fifty percent of the population of the United States. With the exception of some rare organic causes of hypoglycemia, the majority of cases are the direct result of long standing poor dietary habits, especially the tremendous amount of refined carbohydrates consumed each year.

The effects of hypoglycemia span the physical, emotional and spiritual areas of a person's life. The net result is extremely poor quality of life for the person afflicted, and increased stress for family and friends.

The treatment for most hypoglycemic conditions is relatively simple and effective within a three to four month period of time. Treatment is basically nutritional by removing the offending food substances and providing frequent food intake. Vitamins and mineral supplementation is often very important.

Even though the diagnosis is not complicated to make and the treatment usually simple, patients are still too often subjected to ridicule and dismissal by their physicians who say, "Hypoglycemia is a fad disease." Other physicians, either through ignorance or possibly deception, will do an inadequate laboratory test to "prove" that the patient does not have hypoglycemia.

In this book the authors have revealed first-hand

accounts of the suffering caused by hypoglycemia, both from the condition itself and inadequate diagnosis. For the physician and interested lay person there are extensive references to the literature concerning this all too prevalent condition.

Hypoglycemia and its consequences simply need not exist. The authors have provided information which could help eradicate this condition.

Harvey M. Ross, M.D.

Introductions

When I started to write this book, I intended only to relate my experiences with hypoglycemia and those of countless sufferers I had met during radio and TV appearances. But gradually, as I delved deeper, the controversy over the ailment began to emerge, and I found myself trying to sort out truth.

Clearly, the line was drawn between nutritionists, nutrition-minded doctors, and the medical establishment. Previous medical research had not given any conclusive evidence to prove which was right. Was hypoglycemia a real disease, or just a figment of medical imagination? Could it be diagnosed and treated? Why were so many doctors oblivious of its devastating effects?

Little did I realize that my quest to find answers to these questions would end in such a thorough investigation of scientific dogma. The results have been gratifying, however, but are only the beginning. Much more can be done and *has* to be done to insure ourselves and our children a healthy, happy tomorrow.

Maura (Jinny) Zack

When I first saw Jinny Zack as a patient, she had many symptoms of hypoglycemia—extreme fatigue, headaches, crying spells, mental confusion, unnecessary fears and depression. She had just attempted suicide. The anti-hypoglycemia diet she had been on for several months had relieved some of those tensions, but it was only after I discovered and treated the basic cause of her low blood sugar that the symptoms disappeared.

Jinny regained her mental and emotional stability and has spent the last several years helping others who suffer from this much-neglected condition. Her long and harsh experience with hypoglycemia and the prevailing medical attitude toward it was her motivation for writing this book.

Part 1 of *Sugar Isn't Always Sweet* explains how Jinny and other sufferers coped with their condition. It also examines the spiritual effects of this often devastating physical problem. Part 2 will give the reader a better understanding of blood sugar abnormalities. It describes the glucose tolerance test, which is presently the only laboratory method for diagnosing hypoglycemia. This section also contains a comprehensive review of the literature on this subject, both lay and medical, with a clinical description of signs and symptoms. Surprisingly, even to me, the literature establishes reasons for the intense controversy that has raged for years about this ailment. The book gives suggestions for resolving this dispute.

The change in eating habits necessary to control blood sugar levels is discussed in Part 3. Not only must certain dietary changes be made, but frequent snacks are necessary to maintain the proper sugar balance in the blood. In some cases, change of dietary habits alone will control blood sugar. If it does not, the basic cause of the hypoglycemia must be found through further diagnostic tests.

The data presented here lends insight to a medical

problem that has been grossly overlooked and maligned. A powerful compilation of research, *Sugar Isn't Always Sweet* offers synthesis to American medicine and hope to countless hypoglycemic sufferers. It is one of the most authoritative treatises on this subject that I have ever seen. Yet it has a personal touch.

Wilbur D. Currier, M.D.

Contents

Acknowledgments	vi
Foreword	vii
Introductions	ix

Part 1 Living with Hypoglycemia
1 Golden Gate to Life	17
2 Medical Merry-Go-Round	27
3 From Happy Home to Bitter Battleground	37
4 No One Is Immune	47
5 Where Are You, God?	63

Part 2 Understanding Hypoglycemia
6 Stepchild of Medicine	77
7 Millions Suffer Needlessly	95
8 Discovering the Enemy	113

Part 3 Managing Hypoglycemia
9 Eat Right and Save Your Life	125
10 The Hassles in Getting Good Food	139
11 The Rest of the Deadly Don'ts	157
12 RIME, Reason and Rescue	167

Appendix
A Articles on Childhood Hypoglycemia	178
B Less Frequently Named Symptoms of Hypoglycemia	180
C Medical Schools With Courses in Nutrition	181
D Currier Pancakes	182
E Currier Cocktail	183
Notes	184
Bibliography	191

Part 1
Living With Hypoglycemia

1/The Golden Gate

The setting sun was shrouded in the misty San Francisco fog. I planned to sneak onto the Golden Gate bridge in the coming darkness so that no one would guess my intention.

A nearby coffee shop seemed a handy place to sit and wait for dark. I pushed open the door, noticed an empty seat at the long, drab counter, and dragged myself to it.

Not much here I can eat, I thought, glancing indifferently at the menu. The busy waitress smirked when I ordered only a glass of milk.

My gaze turned toward the bridge. As I squinted at the words of a little sign, my heart sank.

"No pedestrians after dark."

For a moment I didn't breathe. That message was devastating. My plan depended upon darkness.

My mind churned frantically, trying to decide what to do next. Turning back never entered my mind, but how could I jump in broad daylight? That prospect was so startling, so defeating to my purpose, that I quickly brushed it aside.

I had planned my death in minute detail, even though I had always believed that suicide is wrong, that God gives life and is the only One who has the right to take it. But for months I had been so confused — one half of my mind insisting, *it's wrong,* the other half taunting, *go ahead, do it, it's the only way out.*

I had considered various methods; my first idea was pills. But what if they didn't work? I didn't want to be a vegetable for the rest of my life. I had thought of racing my car over a cliff. Just a few miles away from my home in Carmel, California, the road follows the coastline where the mountains cascade into the sea. A calculated sharp right turn would send the car careening over the cliff. As it hit either the rocks or the water below, I'd probably be gone. But no, I couldn't do that, either. Houses lined some of those cliffs. I couldn't take the chance of hitting one and killing someone else.

I thought about shooting myself. But where would I get a gun? And how would I learn to use it? I couldn't ask somebody to show me how to shoot myself.

Another idea had crossed my mind: buy some hose, hook it up to the car in the garage, and do it with carbon monoxide. I didn't know how to do that, either, but maybe I could find out. I soon dismissed that notion, too. The children would find me, and I couldn't put them through a traumatic experience like that.

I resolved not to kill myself unless no trace of me could be found. It would be better for my children to go through life wondering what had happened to me, I reasoned, than knowing that I had committed suicide.

The waitress's voice snapped me back to reality.

"Wouldn't you like a piece of pie or something to go with your milk?"

"No, thanks. I'm really not hungry."

Pushing my half empty glass aside, I quickly slid off the stool. I had emptied my purse of all identification, so it didn't take long to find some change. I tossed some coins on the counter and started for the door. Over and over my mind churned, *Daylight or not, I have to do it now. I have to do it now.*

Passing through the crowded restaurant, I noticed the noisy people at their tables. Everybody seemed happy and cheerful, glad to be alive; I just wanted to die. They all seemed so contented and full of fun; I was full of . . . nothing. Thank God, it would soon be over.

The door opened easily, and I stepped into the chilly fog. The top of the Golden Gate rising out of the mist probably looked beautiful to everyone in the coffee shop; to me, it was just a cold, steel means to an end.

As I headed for the bridge, my pace slowed. The thoughts pounding inside my brain dulled the noise of the automobiles whizzing by a few feet away. How I wished I were alone! How I wished that it were dark and there weren't so many cars! Jumping would be so much easier.

Through the open framework of the bridge, I could see the concrete pilings beneath. Jumping would have been a snap at this point, but I couldn't bear the thought of splattering on solid rock. My goal became the middle of the span where there was nothing but water below.

Suddenly, my life started to pass before my eyes, just as if I were drowning . . . .

Once I had been a joyful child, then a content, though rather insecure, teenager. After my marriage at thirty-two, I was a happy, calm, positive, loving wife and mother and an efficient business partner to my husband, Jay.

After adopting two babies, Jerry and Julie, I gave birth to Jimmy. Then we moved from Los Angeles to one of the most beautiful spots on Earth — Carmel-by-the-Sea. The motel we bought there was really a group of Hansel-and-Gretel-like cottages, which had a sign in front that read, "Carmel Cottage Court." We deleted the "Court," making it "Carmel Cottages," and added "Little Bit of Heaven."

My year of business college and several years as a secretary and public stenographer paid off. I was able to handle the motel office alone. Greeting the guests, showing them to their cottages, and verbally introducing them to our lovely area awakened memories of my two years as an airline hostess during World War II.

Our Little Bit of Heaven, plus an affectionate, loving husband and three beautiful children, should have made my life fulfilling. But they didn't.

I remembered standing in front of the kitchen sink one night and saying aloud to myself: *I don't have any joy in my life anymore. I hate these cottages, I hate the kids, I hate Jay, and I hate myself.*

I had become an unhappy, negative, whining, cantankerous, nervous, middle-aged woman who was usually too tired to enjoy anything.

During my annual physical, I would ask the doctor, "Why am I always so tired?" Since all my tests were normal, his answer was always the same. "Well, Jinny, you have a hard job, three small children to take care of, and you're past forty years old. It's just emotional." After about five years of that, he decided that I could be starting through the menopause, and prescribed some hormones.

They didn't help a bit.

We hired managers for the motel, bought a house

and moved. For six years I had often complained jokingly of working twenty-eight hours a day, eight days a week. At least, that's what my schedule usually seemed like. I presumed that getting away from the pressures of the motel would make me feel better.

It didn't.

I soon became obsessed with a new idea. Since our doctor believed there was nothing wrong with me, Jay must be the cause of all my problems. I'll never forget the night I told him, "I just can't live like this anymore. The only solution I can see is for you to leave." The next day he was gone; shocked, but gone.

But this maneuver didn't work either. My tiredness turned into exhaustion, and the unhappiness into depression. Eventually, some friends suggested that I try taking vitamins. Determined to take the right ones, I went to a health food store where the woman behind the counter listened intently as I described how I had been feeling.

"Sounds like you have hypoglycemia," she offered.

"HypoWHAT?" I asked.

"Another name for it is low blood sugar," she answered.

"I couldn't have anything wrong with my blood. My doctor just gave me all these exhaustive tests...."

"Did you have a six-hour glucose tolerance test?"

"What's that?"

"If you'd had one, you'd remember it," she smiled knowingly. "You fast all night and go into the lab the first thing in the morning. They give you some glucose to drink, then take samples of your blood for the next six hours."

"I never had anything like that."

"Then you couldn't know whether you have hypoglycemia or not. That's the only test there is for it. I suggest you have your doctor give you one."

The clerk was right, and ever since that time, I've called her my lifesaver. When the doctor saw the results of the glucose tolerance test, he admitted that the blood sugar was too low, but confessed that he knew nothing about the condition. He knew of no other physician to recommend, so I was left on my own to find help. At least now I knew that hypoglycemia, not Jay, was my problem.

But the conflicting opinions of the sixteen doctors I visited during the next two years were almost harder to bear than the symptoms of my illness, and I despaired of ever getting help.

A first century writer described my feelings exactly when he said: "The fear of death is more to be dreaded than death itself." It was in such a state of mind that I found myself on the Golden Gate bridge.

As I stood there lost in my thoughts, a glance over the rigid metal railing assured me that only icy-cold black water lay below. But that railing! Although barely chest high, it seemed insurmountable. How would I ever climb onto it, let alone go over? There seemed to be no place for my foot to get leverage. All I could see were metal bars criss-crossing one another to form a sturdy barrier.

As my hands grabbed the frigid metal, I wondered, *Can I pull myself up that high?* I stared down at the dark water. My legs felt rubbery. *It's so far down! The water must be freezing! What if I don't die?* Something inside seemed to be pulling me away. *If I jump, everything will be over.*

I forced my hands back on the railing and peered below. The whitecaps looked like tiny ribbons. *What if the current doesn't wash me out to sea? What if it carries me to shore instead?* I had envisioned that my body would never be found and that the children would never know what happened to me. *If they find me, the kids will have to carry that shame around with them all their lives.*

I started to pull away again, then changed my mind. *They'll get over it, but I won't ever get over feeling like this. There's no one to help me. I just can't go on.*

I looked around. *What if someone sees me—and stops me?*

California law requires that anyone who attempts suicide must be placed under observation for seventy-two hours. If I were stopped and sent to a mental hospital as a would-be-suicide, everybody—the kids, their friends, everybody—would know it.

I couldn't afford to be caught in the act. I would have to be very cautious. Could I do it? Could I pull myself over the railing without attracting attention?

I tensed my muscles again, but couldn't lift myself more than a few inches off the ground. I tried changing my grip and my stance, but nothing seemed to help. I simply wasn't strong enough to climb over that railing.

In defeat, I stopped struggling, and leaned limply against the rail, watching the shadows lengthen across the water. Then, reluctantly, I pushed myself away, turned around slowly, and trudged back the way I had come.

I had failed again.

Half stumbling past the coffee shop, I felt as though a hundred condemning eyes were staring at me. In truth, I suppose no one had even noticed me leave. Plodding toward the Presidio, I knew that my legs would never make the long hike back to my motel room.

"Do you know if there's a bus stop close?" I questioned a tall, grey-haired man strolling toward me.

"Just keep going straight. It's at the end of the next block," he smiled, pointing toward it.

"Do you know if it goes to the gate on Lombard Street?"

"Yes, it does," he nodded.

A bus arrived soon, and I stepped gingerly aboard, almost expecting someone to reprimand me for what I had just tried to do. Although I didn't enjoy the long, winding ride through the Presidio, I couldn't help but notice the gorgeous trees and flowers, the rolling hills, and the ocean. The water looked beautiful, not at all like that ominous, cold bay below the bridge.

Stepping off the bus, I saw a liquor store. With a sigh, I turned toward it. *Maybe a drink would cheer me up!*

As I opened the door, a pleasant voice greeted, "May I help you, Miss?"

How I wished some of his cheerfulness would rub off on me.

"I don't know," I answered. "I just think maybe a drink would cheer me up."

"What do you like?"

"Nothing especially. I don't like the taste of liquor, but it's all right if it's with something sweet."

"How about one of these already mixed drinks? A mai-tai or a daquiri, maybe?"

"The mai-tai would be okay, I guess."

I paid him and headed for the motel. Once in my room, I assured myself that my depression would end soon. As I took a glass from the dresser top, my eyes lit on the TV next to it. I turned the knob. Several shows and a few drinks later, I was more depressed than ever.

As the TV droned on, the thoughts once again raced through my brain.

How can I get out of this? Why can't somebody help me? What'll I do?

Suddenly, I felt prompted to call the Suicide Prevention Bureau. I dialed their number, and within minutes someone was tapping softly at the door. I glanced at my watch. It was past midnight.

Opening the door, I was surprised to see a short, bearded man in a sports jacket and a Roman collar. *What do I need with a minister?* I wondered. I fumed inwardly, but invited him in and turned off the television.

When he stepped into that room, I'm sure he had no idea that, when he left, the sun would be coming up over San Francisco, or that he would have heard me recount all the agonies that had led me to this point.

Neither did I realize that it would be the dawning of a new life for me.

2/The Medical Merry-Go-Round

"**H**i. I'm Bob Anderson from Suicide Prevention."

"Hello. I'm Jinny Zack," I muttered. "Thanks a lot for coming."

I motioned him toward the only chair in the small, dimly lit room. He sat down, and I threw myself across the bed.

"Why do I have to go on living?" I blurted. "I'm no good to anybody. Can't do anything right. Couldn't even jump off the bridge. Can't even take care of *myself*, let alone the kids, the house, the car, the grocery shopping, the cooking, the...."

"Ho-o-o-ld on! Why don't you just be quiet for a minute? Try to relax, then maybe you can sit up and tell me all about it."

Bob sat quietly, waiting for me to calm down. When I finally rolled over, sat up and leaned against the headboard, he smiled patiently.

"Now, do you feel like talking?"

"I guess so."

"You said a moment ago that you can't take care of yourself. What did you mean by that?"

"Well, I have hypoglycemia, and I'm tired all the time, and I have these awful headaches, and I can't concentrate ... can't even think straight most of the time, and I get so depressed I can't stand it. I can't cope with the kids, can't cope with anything anymore...."

"Have you seen a psychiatrist?"

I sighed. How many times had I heard that before? "Please don't tell me it's all in my mind. I know it isn't. Hypoglycemia *can* do all that to you. They don't know what to do for it. I *know* it's what caused my husband and me to separate."

"How could hypoglycemia cause that?" he asked skeptically.

"Lots of husbands can't stand to be around their wives when they have hypoglycemia, but Jay was willing to wait for me to get better...."

"Why didn't he?"

I started to cry. Bob stood, pulled his chair closer, then sat back quietly.

"I'm sorry, but I have to know these things if I'm going to help you."

"I know, but it hurts to talk about it. You see, he has a sister-in-law who's a nurse. When she was visiting us one time, Jay and I told her about my problems, and that the doctor couldn't find anything wrong with me except that I was probably going through the menopause. She said it takes some women six or seven years to get through the change of life. I know that Jay was willing to wait, but I got to the point where I couldn't stand things any more. I didn't get any better with the hormones the doctor gave me, so I just knew Jay was the cause of everything. I couldn't cope with him. I told him to leave, and he did."

"Well, if you thought your husband was the cause of all your troubles, why are you blaming it all on hypoglycemia? A lot of doctors say it's just a fad, you know."

I stiffened. "A fad? How could they call it that? Do you know what a fad is?"

"Well, it's a ... well, you know ... it's a ... it's the fashion ... the style ... it's all the rage, everybody has it."

"The fashion? Can you imagine me wanting to go through all this just because it's the fashion or because everybody else has it? Whoever calls it that doesn't know what they're talking about! Besides, I still had a problem even *after* Jay left!"

I jumped up and started pacing the floor angrily.

"That's what makes me so mad about this whole thing. People who say things like that just don't know! Hypoglycemia is a real physical condition. The blood sugar is so abnormal that it causes all kinds of problems. Those who say it's just a fad have never suffered like I have. They've never been so depressed they didn't want to live. Do *you* think I tried to jump off the bridge just because it's the rage and everybody's doing it?"

"Well, hardly."

"There've been lots of times when I felt like I was going crazy, but I wouldn't be crazy enough to choose to have hypoglycemia just so I'd be in style. Besides, I've had glucose tolerance tests which *show* that I have it. How could a doctor or anybody else argue with a clinical test?"

"Okay, but I was just telling you what I've heard."

"Well, what you heard is wrong. I have lots of symptoms...."

Calming down, I began to describe my condition more rationally. I was tired all the time. I'd go to bed at seven o'clock and get eight or ten hours sleep and be just as tired when I woke up as when I went to bed. Then I started waking up after about four hours and couldn't get back to sleep – a typical sleep pattern for a hypoglycemic, I later found out.

Then my whole personality started to change. I didn't know it was happening, but Jay did. I'll never forget one thing he said to me one day when we were having a big argument: "You're just not the same sweet girl I married. I don't know what's wrong with you – you're up one day and down the next."

It was as if my emotions had gone haywire. One day I'd be my usual self, and the next I'd be a crab. Then I became grouchy most of the time. I thought it was all his fault; he thought it was mine.

"Did you ever see a marriage counselor?" Bob inquired.

"Yes, but that didn't help."

I also explained that our family doctor tried to help, but his tests showed that there was nothing wrong with me. I knew that there was, and so did my friends. One suggested I take some vitamins. I went to a health food store so I'd be sure to get the right ones, and the girl there recognized my hypoglycemia right away.

Bob was incredulous when I told him that she had identified the problem when the doctor hadn't. "How could *that* be?" he asked.

"I s'pose it's because hypoglycemia is a nutritional problem and that's what health food stores are all about. My doctor admitted he didn't know anything about it. Lots of doctors don't know much about it because they don't study nutrition.

"Anyway," I continued, "the clerk suggested I have a glucose tolerance test, and she also gave me a book to read – *Low Blood Sugar and You*. I asked my doctor to give me a test. He did, and it was positive. But since he couldn't help me, I was determined to find one who could. I asked around, and there wasn't a nutrition doctor on the whole Monterey Peninsula."

Restlessly, I related how a neighbor had told me that her husband had just learned that he had hypoglycemia. I called his doctor right away. He wasn't accepting any new patients, but agreed to see me about my low blood sugar. He looked at my test results and confirmed that I had it, and that the diet I had gotten from *Low Blood Sugar and You* looked okay. He suggested, though, that to make sure, I write to the Hypoglycemia Foundation and get their diet.

I did, and the diet was the same. But I still didn't feel right about treating myself from books, and kept on looking for a doctor who could help. About that time, someone told me about a psychiatrist who was familiar with hypoglycemia, and I went to him. He confirmed that my test was positive and advised me that as long as I stayed on the diet I would be all right. I didn't know then that psychiatrists are also M.D.s. I wasn't getting better, so I continued my search.

Hearing about a doctor in Anaheim, about four hundred miles away, I flew down to see him. He gave me a battery of examinations, including another glucose tolerance test. He discovered that my adrenal glands weren't functioning properly which, he said, was causing the hypoglycemia. He prescribed injections of adrenal cortical extract. ACE, as it is commonly called, is an extract from the glands of animals and helps restore our adrenals.

But my own doctor wasn't familiar with it and at first refused to give me the injections. I begged Dr. Brown to call the specialist in Anaheim, and he did. After he learned what ACE was and how it was going to help me, he agreed to give me the shots.

In a few weeks, with the injections and the diet, I began to feel like a human being again—no more headaches, no more crying spells, and the depression was almost gone. Unfortunately, Dr. Brown left for a conference and asked his new associate to give me the shots. He did, but let me know that he didn't believe in them and that he didn't think they were going to do me any good.

Those negative feelings must have been transferred to Dr. Brown, because shortly after he returned, he announced that he wasn't going to give them to me anymore. He didn't say why, but I always suspected that his associate had turned him against them.

I didn't know what to do. Without the ACE, even though I strictly followed my diet, my condition deteriorated again and my emotional problems returned. One day I felt so awful that I just stood in the living room and banged my head against the wall. When that began to hurt, I stood behind a big over-stuffed chair and pounded my head against it as hard as I could.

Finally, I began to realize that I hadn't felt this way when I was getting ACE. In desperation, I called a doctor friend who was with the county health department and asked him to give me the injections. He said he couldn't give them to me at his office because they were not routine treatment for the county, and he couldn't give them to me privately because that would be moonlighting and the county frowned on that.

He did suggest three doctors whom he thought might give to me. The first one I visited refused, explaining that he didn't know anything about ACE. The second said that he was giving it to a woman who was visiting in the area. He hadn't heard of it previously, but her family doctor recommended it, and it seemed to be helping her.

I was elated, but wouldn't have been had I known

what was going to happen next. He set up an appointment for Saturday. On Thursday morning I woke up crying. I sobbed all day, most of the night, and all day Friday.

Saturday morning I wept until it was time to leave the house. I managed to stop while I drove to the doctor's office, but as soon as I arrived, I started in again. As the doctor stepped into the examining room where I was, a puzzled look crept across his face.

"What's the matter, Mrs. Zack?" he asked kindly.

"I don't know. I've been crying for two days and can't stop," I whimpered, fighting hard to control myself.

"Would you mind if I put you in the hospital?"

"Oh, I wish you would."

I had been dreaming for months of having lots of doctors hovering over me in a hospital trying to come up with *the* answer. But his next question shocked me. "Do you know any psychiatrists in the area?"

"Why? Are you going to put me in the psychiatric ward?"

When he answered "yes," I didn't have the strength or the reasoning power to argue with him. I called Jay from the doctor's office, told him I was going to the hospital, and asked him to stay with the children for a few days.

After I had been there for five days—with no crying spells or symptoms of any kind—the doctor gave me another glucose tolerance test. When he saw the results, he announced cheerfully:

"You can go home now, Mrs. Zack. You don't have to worry about hypoglycemia any more. Your test is normal."

When I saw the results, I knew it wasn't normal. There was a drop in my sugar level from 183 to 114 in the second hour. By the fourth hour, it had fallen to 58. The doctor was wrong, but it was useless to argue.

As soon as he left my room, I called three other doctors who were knowledgeable about hypoglycemia. Each, upon hearing the results of that test, agreed that I did, indeed, have hypoglycemia.

After I left the hospital, a friend suggested that I see an internist. She reasoned that a specialist in internal medicine would know more about my condition than most general practitioners. She suggested Dr. John McCarthy, and I made an appointment with him.

He studied the results of my latest test, peered at me over his glasses, and shook his head. "Mrs. Zack, you don't have hypoglycemia. This test is normal."

He mumbled something about seeing a psychiatrist as I stumbled toward the door. By this time I was bewildered and discouraged, but had to keep looking for a doctor who could help me. Hearing of a nutrition-minded doctor in Santa Cruz, about fifty miles from home, I went to see him. He gave me another glucose tolerance test, which was also positive, and put me on a different diet, promising that I would start to feel better soon.

When I didn't, he referred me to an endocrinologist at Stanford University Medical Center. I was thrilled! If anybody would have the answer, it would be he. But what a disappointment when I saw him. That specialist studied my latest test while I described my symptoms. His diagnosis was all too familiar.

"Mrs. Zack, you don't have hypoglycemia. Besides, tiredness isn't a symptom of it. Neither are headaches or depression. Adrenal cortical extract is absolutely no good, and I'm going on a crusade to put the Hypoglycemia Foundation out of business because they recommend its use."

I couldn't leave his office fast enough. On the way home I phoned the doctor who had sent me there. His only comment was, "Well, there are a lot of doctors who aren't knowledgeable about hypoglycemia."

Bob listened intently to everything I said, but was still not convinced. "Jinny," he said emphatically, "I still don't see how a physical condition can do all this to you. It must be something else. Tell me about your husband."

3/From Happy Home to Bitter Battleground

In those early hours of the morning, I wearily began to recount the events that had led up to Jay's leaving. "Jay was one of the most affectionate men in the world. He must have told me fifty times a day that he loved me, and he was always kissing me. We held hands every place we went, even in church. And every time we were in the car, I sat close to him so that he could put his arm around me. He always drove with one hand and even received a traffic ticket for it once. Whenever we bought a new car, he wouldn't even consider one with bucket seats because I couldn't sit close to him."

Bob eyed me curiously as I recounted happier days.

"We were married on a Saturday, and the next week he brought me a present for our 'anniversary'." I smiled as I remembered. "For five years after that, we gave each other a present every Saturday."

"That's really something. How long were you married?"

"Sixteen years."

"But you stopped the presents after five years?"

"*I* did, yes."

"Why?"

"I just plain ran out of ideas. Jay didn't drink or smoke; he didn't play golf or have any other hobbies. I got tired of buying candy bars and Life Savers and tools and trying to think of something to get him."

"What did he think about stopping the presents?"

"Well, he didn't want to, but he stopped when I did. It didn't seem to make any difference."

"Are you sure?"

"Why do you ask that?"

"I'm trying to uncover something that may have started a change in your relationship."

"Well, I don't think it started then. Our troubles didn't really start until after we had been in Carmel a few years."

I explained to Bob that when we had come to Carmel eight years before, it was truly our Little Bit of Heaven—for nearly four years. Gradually, though, things changed. I couldn't talk to Jay without arguing, and the children began to get on my nerves. I enjoyed long talks with friends, even strangers, and spent hours conversing with guests. Being around other people was enjoyable, but I couldn't stand my own family.

Later, I remembered something my mother had told me when I was young. She was a nurse, and while working in a mental hospital had noticed that when people were losing their minds, they always turned on their loved ones first.

That's what had been happening to me. My brain was being starved, and I was acting like a person who was losing her mind. I, too, was turning on those I loved, especially my husband.

Those days weren't pleasant to recall. Jay often said he loved me, but I didn't believe he showed it. For fourteen years he had treated me as though I were the most important thing in his life. Suddenly, he wouldn't talk to me. We had always made big decisions together, and had similar ideas about raising children.

Thinking he had changed, I fought with him about the kids, arguing about every little thing. He wouldn't confide in me, and took care of a lot of matters without even consulting me. Because of this, I didn't think he loved me anymore.

"I remember one time," I confided to Bob, "he had to go on a business trip for a few days. I decided to surprise him by redecorating one of the cottages while he was gone. I shopped and shopped for curtains, even went to a town twenty miles away to buy the material. I had a bedspread made to match, and painted the room myself. I really worked hard.

"When he got back, the first thing I did was show him my handiwork. I thought he'd really like it. Instead, he wanted to know why I had done it — said it looked terrible."

"Sounds like he had become a pretty insensitive fellow," Bob injected.

"I was crushed. And the more I brooded about it, the worse I felt. I hated to show people that cottage. It always reminded me of what he had said.

"Another time when one of our maids hurt herself, we had to see a lawyer and go before the Workman's Compensation Board. I prepared a lot of records, and we talked about it for days, discussing what we were going to say when we arrived. On the morning we were supposed to appear, I was dressed and ready to leave when, without warning, Jay informed me he was going by himself. Shocked, I cried and pleaded with him to let me go, but he wouldn't."

"Did he say why?" Bob wondered.

"No. And there was something else that used to

bother me. Jay started all kinds of projects, but it seemed to me he never finished them. That made me furious."

"Maybe you had a right to be," Bob offered. "It sounds like Jay *had* changed a lot. Maybe he's had more to do with your problems than you think. I still can't see how hypoglycemia could cause all those emotions, but I *can* understand the feelings you must have had because of his actions."

Bob droned on and on about my relationship with Jay. Words like "domineering," "demanding," and "overbearing," pierced me like sharp needles. He said that I was oppressed and had not been allowed to express my feelings. Maybe he was right! As I thought about the time that Jay wouldn't let me go to the board meeting, my anger flashed. So Jay *was* the cause of all my troubles!

Suddenly, my mind wandered back over the events that led to that desperate flight to San Francisco. For weeks I had been planning a vacation with the children, and they were so excited about it. The day before, though, I had awakened in the wee hours, thinking about that trip. How could I ever stand those tours through the Grand Canyon, the Petrified Forest, Carlsbad Caverns, and all the other places they wanted to go? Where would I get the energy when I didn't even have enough to get up and get dressed most days? How could I endure all those visits to old friends and relatives?

People I loved and had been wanting the children to meet now loomed as monsters in the shadow of the night. I just couldn't do it. I couldn't take that trip. But how could I get out of it?

I buried my head in the pillow. *If only my brain would slow down! O God, please help me. Let me go to sleep. I want to go to sleep and never wake up.*

But sleep wouldn't come. I lay there wide awake. Suddenly, a solution flashed before me. *Why don't I jump off the Golden Gate Bridge?* Quickly, the idea penetrated into my innermost being. *I'll fly to San Francisco, go to the bridge, wait until after dark, then jump.*

Not many who had done it had survived, so the chances of success were good. Once my body hit the water, the current would take it out to sea, and it would never be found.

My plans began to form rapidly. I'd fly to Los Angeles first and get rid of all my identification so that if my body were found it could never be traced. My fingerprints had never been registered, so they couldn't be checked. If I did it right, no one would ever know. The decision was made!

I hadn't bounced out of bed with such vigor in years! Dressing took five, maybe ten, minutes. A phone call to Jay from Los Angeles would alert him to take care of the kids. I dashed to the car to begin my last trip.

When I reached the airport, however, I was surprised to find that it didn't open until six o'clock. It was then only four. Driving around the peninsula those next two hours, I watched the sun come up and the lights of the city go out. The waters of the bay began to sparkle, and stately pines and windswept cypress trees took shape in the early dawn. . . .

Lost in my thoughts, I wasn't aware of the silence that had settled in the motel room.

"Jinny." Bob's quiet voice brought me back to the present. "I think you'll feel much better about things once you get used to having Jay out of your life. He's gone, and you have to face it. Besides, it sounds to me as if there might have been something else—maybe another woman?"

Before I even had a chance to reflect on his words, Bob's callbox beeped, and he rose to his feet.

"I have to go now," he spoke hurriedly. "I hope I've helped you some...."

I shrugged wearily. As he stepped toward the door, I broke my long silence. "Thanks for coming, Bob. I really appreciate it."

He opened the door just as daylight was creeping over the San Francisco skyline. I glanced at my watch. We had been talking for almost six hours.

Moments later I called a cab and arrived at the airport just in time to catch a plane home. Once the plane was in the air, I tried to sleep, but couldn't. I kept thinking about what Bob had said.

Did Jay cause all my problems? What about all those doctors who said hypoglycemia was causing them? How could I let Bob sway me so easily? He did sound convincing. Maybe he's right. Maybe it is Jay's fault. But I know he's wrong about one thing—another woman. We were almost never apart until I sent him away.

Suddenly I thought of the note which Jay had written four days after he left. That would prove that he wasn't the monster Bob made him out to be. I determined to look for it as soon as I reached home.

A pleasant voice on the inter-com interrupted my thoughts.

"We will be landing at the Monterey airport in just a few minutes. Please fasten your seatbelts. Thank you for flying United. We hope that you've enjoyed your flight and that you'll fly with us again soon...."

I looked at the clock as I strolled into the terminal. *7:30—much too early to get the kids up.* I headed toward my car which I had left in the parking lot the morning before. I rode around for awhile, rehearsing what to tell the kids about my disappearance for a day. The same gorgeous peninsula seemed to welcome me back. The beautiful sunrise of the day before was hidden by the more typical June fog, but I didn't mind. When I felt good, I loved the fog.

Once home, I found Jay's note in my desk drawer. In his familiar hand, it read:

To my dear wife:
May God bless you in your new life, and keep you and the children safe. Please keep my children unspoiled, as they are, and hold a good memory of me for them. I can ask no more, knowing they couldn't find a better mother anywhere.

I will treasure always the happy years, so many of them, and hold them dear to my heart. The wonderful hours when I held you so close, will never fade. Your lips on mine, a memory. Always, always, always, will I remember you as you were, gentle, kind and loving. You, you, were my inspiration when my heart was only for you. A thousand times you alone held me up, when down was so far. I guess you didn't know how great was your strength and how deep was my love. It will be long before I forget. Although my face is calm, my heart is breaking. For all the hurt I have given you, forgive me and forget me not.

These words I cannot say to you, but from the heart, with eyes full of tears, they are yours. Go forward, my love, find your new life, be brave and strong, lift up your eyes to the sun and stars and go forward. Do not tarry alone. Go to the cities, seek and find a helpmate worthy of you. You deserve the best! May he find in you the answer to his dreams, and you in him. I wish him every happiness, and if he but finds a tenth of what I had, he will be rewarded beyond his fondest hopes.

Always, your loving Jay

After rereading it, I was certain once more that hypoglycemia, not Jay, was my biggest problem.

That note had been written in April. On Mother's Day, I received one red rose in a crystal vase from Jay. Later, I found out that one red rose means, "I love you." I enjoyed getting it, just as I would have enjoyed a gift from anyone. I didn't know then about all the tears Jay had shed the first few months after he had left. Friends told me about that later. Nor did I have knowledge of the many times he had told the children he wanted to come home. I don't remember that he ever asked me about coming home.

On my birthday in August, on Thanksgiving, Christmas, and Easter the following year, I received a single red rose. His love may have been deep, but I've often thought it must have soon been over-shadowed by the hurt of my rejection. He never did understand my sickness. He, too, thought it was all in my mind, and that I'd never again be the "same sweet girl" that he married.

By the time I received adequate help and returned to my former self, it was too late. Jay had been gone more than two years. He had met somebody else and was planning to be married.

Years later I understood Jay's attitude. It was true; before he left, I had become a different person. I disagreed with him about everything, especially the children. I flew into a rage at the slightest provocation. That's probably why he quit talking to me. My redecorating ideas could have been pretty awful, and I could easily see why he didn't want me to go to that board hearing. He never knew what I was going to say or how I was going to act. He couldn't trust me in such an important meeting.

Jay hadn't changed; I had.

Although from the beginning our marriage was not perfect, we loved each other deeply. I'm convinced today that I could have put things in proper perspective had my brain been functioning properly....

I woke the children up as usual that morning after I came home from San Francisco. After a brief explanation about my absence, it didn't take us long to finish packing for our trip, eat some breakfast, and get started. Our first stop was Riverside to see their cousins. From there I called Dr. Wilbur Currier in Pasadena and made an appointment. I had met Dr. Currier once when he was in Carmel, had talked to him frequently on the phone, but had never seen him as a patient. I made an appointment to see him the next day.

After my initial examination, he gave me three tests that none of the other doctors had given me. One was a diganex-blue test. Swallowing the capsule containing the blue dye for that test took only a few seconds. Getting the report back from the lab took a few days. Such a simple process! Yet, it held the answer!

Malabsorption was the basic cause of all my difficulties. I didn't have enough hydrochloric acid in my stomach to absorb the food I ate, and this was upsetting my whole system.

Today, that test has been replaced by the Heidelberg Gastric Analysis which involves swallowing a capsule containing a radio sending mechanism. The contents of the stomach are transmitted to a computer, which gives a complete print-out—much more sophisticated and accurate.

He advised me to stay on the diet, but to start taking a digestive enzyme containing hydrochloric acid. Two tablets with every meal built up the acid that my stomach was not producing, and soon I was on the road to recovery.

Exactly what was this thief that robbed me of happiness and made my home a battleground? How could there be such contradictory views over a real physical ailment? Why would a doctor at one of the most prestigious medical centers in the country disagree with so many I had seen and others whose books I had been reading?

I determined to find the answers.

4/No One Is Immune

Shortly after I became free from my physical and emotional torture, I resolved to do anything possible to keep my life-shattering experience from happening to others. But for this task, I needed to know much more than I did.

I began to study every book and magazine article about low blood sugar that I could find. About this time, I also discovered two helpful research tools at the public library, *Books in Print* and *Reader's Guide to Periodicals*. *Books in Print* lists all the books written in the United States since 1948, both by subject and author. The periodical guide identifies articles in popular magazines by subject since 1896. I bought most of the listed books for future reference, and made copies of pertinent magazine articles.

Realizing I also needed information from a more scientific source, I made several trips to Lane Medical Library at Stanford. The stack of medical journal articles I copied there was seven inches high. In addition, I interviewed nutritionists, doctors and others in the medical profession.

I shared my findings on radio and television talk

shows and organized study clubs to exchange information and experiences. Many people who were diagnosed as being hypoglycemic or who suspected they might have it attended those meetings.

My exhaustive studies confirmed to me that hypoglycemia is a medical and nutritional problem and that my sufferings were not caused by an emotional condition. I also learned that this condition occurs in various degrees. To one person it might be just a bother, something to put up with, while to another it may become as insidious and destructive as it was for me.

I recall the heartbreaking events surrounding the death of Patty, a young woman who attended one of the study club meetings which I conducted. I had just finished discussing hypoglycemia on a small FM radio station in Carmel. During the show the host asked me how it had affected my life. Among the things I mentioned was the break-up of my marriage.

We were not even out of the recording studio when the phone rang and, to my surprise, it was for me. A man's voice said, "My name is Art. I just heard you talking about hypoglycemia on the radio. My girl friend has it, and about three weeks ago she made me move out of our apartment. After listening to you, I think maybe it's her hypoglycemia that broke us up."

"It can do that," I replied. "I know that's what happened to my marriage."

"Is there anything I can do?" he asked pleadingly. "I love her and want to go back. Do you think it would do any good for you to talk to her?"

My heart went out to this stranger whose voice sounded as if I were his only hope. As I quickly jotted down his girl friend's name and phone number, her name sounded familiar. Later, I checked my records from the study club. Sure enough, she had attended a meeting.

A few days after his call, Art was arrested on suspicion of her murder. When I read the account of the killing in the paper, I thought: *I'll bet there's some connection between that crime and hypoglycemia. I'd like to talk to Art again and find out what happened.* I wanted to see him, but was apprehensive about going to a jail. Nevertheless, I found out just where to go and when I could see him.

On the way to Salinas, the county seat, I wondered what I was getting myself into. I had no idea what Art was like, nor could I picture what it would be like to visit him in jail. But I wanted to know him and learn the details of that crime. From the tone of the article I suspected that Art may have been having a hypoglycemic "attack" when he killed Patty.

The long, dim waiting room was crowded with people biding their time until they could have their weekly twenty-minute visit with a friend or relative. Along one wall was a bench where I sat and tried to read. The opposite wall consisted of a series of glass panels just wide and tall enough for someone to stand in front and be on a level with the person behind the glass. Hanging on the wall at the side of each "booth" was a telephone used to talk to the prisoner.

Art was in jail for the first time in his life. The only other encounter he had had with the law was when he was a teenager — he and some other boys had stolen a car and gone for a "joy ride." Now he was in jail for murder.

I soon found out that he could hardly believe it himself. He had spent hours since he had been there trying to figure out what had made him do such a horrible thing. He loved Patty and had gone to her apartment to see if she would take him back. I listened intently as he recounted the events of that afternoon:

"We were on the couch in the living room when I told her I wanted to come back. Everything was quiet and peaceful. I hadn't eaten anything for three days, but I'd stopped to see some friends late that morning and had a few drinks with them. I felt nice and relaxed. I was sitting there with my feet up on the coffee table and my hands in my pants pocket. She seemed to be in one of her better moods until I said something that made her angry.

"I never knew how she was going to react about anything, but I was really surprised when she reached over and slapped my face."

"That's a typical hypoglycemic's reaction—getting upset over nothing," I interjected.

Art said he jumped up, pulled his hands out of his pockets, and saw his pen knife drop to the floor.

"I always carried that knife to tamp my pipe, and I had been fiddling with it while we were talking. When I jumped up off the couch, I stood rubbing my cheek, wondering why in the world she had slapped me. Suddenly, she stooped down, picked up the knife, opened it, and started slashing at me."

All Art could remember was that he tried to get the knife away from her. But the next thing he knew, she was lying on the floor in a pool of blood. He thought she was dead. Panic stricken, he stabbed himself, hoping to die too. Failing that, he called the police, told them what he had done, and fled.

"I decided to go to Los Angeles, so I jumped in the car and drove toward the freeway," he sighed. "But before I got to the on-ramp I spotted a phone and decided to call Mom."

Persuaded by his step-father to return to Patty's apartment, he arrived there shortly after the police, and was placed under arrest.

"Have you ever been tested for hypoglycemia?" I asked calmly.

"No. Why?" Art's puzzled face peered intently through the glass window.

"Because what you did that afternoon and the three days before 'shout' hypoglycemia to me. A hypoglycemic usually has to eat every two or three hours. Going without food for three days and then having a few drinks is almost sure to cause a hypoglycemic reaction. It seems to me that there might have been two persons with low blood sugar in that room. Has anyone ever suggested that you might have hypoglycemia?"

Art shook his head thoughtfully.

"No, but when I was in the service I was in sick-bay all the time. I was never real sick, but there was always something wrong. I had a lot of headaches, was nervous, and a lot of times I couldn't even think straight. I think they labeled me a neurotic. Anyway, I eventually got a medical discharge."

"Art, all those things you mentioned can be symptoms of hypoglycemia. When your blood sugar is low," I explained, "your brain is being starved. Blood sugar is the only source of energy to the brain, so if your brain isn't being fed, anything can happen. Do you think you could get a glucose tolerance test and find out if you have it?"

"I could ask my lawyer, but I probably won't see him for a while. Could you call him for me?"

Our visiting time was up, and I had to leave. A few moments later, I was dialing the attorney from a phone booth. He was anxious to do anything to help Art's case, and on Monday morning he arranged for the test. The results confirmed my suspicions.

Did hypoglycemia contribute to that crime? Did Art know what he was doing when he plunged that knife into Patty's body? Dr. Currier testified at his trial that he believed Art's brain was so starved at the time of the crime that he lost complete control and did something that he would *never* have done otherwise.

I remember that when my blood sugar condition was at its worst, I did many things that I wouldn't ordinarily have done: things like slapping the children, which is strictly against my principles; crying my head off for nothing; saying hurtful things. Had I been attacked, as Art was, who knows what I may have done.

I distinctly remember two times when my children came into the kitchen where I was trying to prepare dinner. They started arguing and fighting. I was feeling so ill at the time that I could see myself going to a drawer, grabbing the biggest butcher knife there and going after them. Why I didn't do it, and why Art did, is simple to explain. I had not been attacked. And I doubt if my blood sugar was ever as abnormal as his probably was at the time he stabbed Patty.

One of the worst things a hypoglycemic can do is to go without food. Because Art hadn't eaten for three days, he was like a ticking time bomb. And when Patty lunged at him, he exploded.

Since it had been proven that Art had hypoglycemia when he was tested a few weeks after the crime, Dr. Currier was sure that he had had it on the day of the murder. However, four court-appointed psychiatrists who, by their own admission, knew nothing about hypoglycemia disagreed with Dr. Currier. They argued that Art was sane at the time of the incident and was responsible for what he had done. The judge chose to believe the psychiatrists, and Art was convicted of second-degree murder. He served four years of a five-year-to-life sentence in state prison, receiving one year off for good behavior. Now that he is out of prison and aware that he has hypoglycemia, he must take responsibility for controlling it to prevent future outbursts.

Shortly after Art's trial, an article about a case in England came to my attention. There, a doctor's wife, a woman of excellent character, drove her car through a small town late at night. Later a bicyclist was found dead in the road along which she had passed, and parts of her car were discovered at the scene.

Mrs. D., however, denied any knowledge of the accident and, when confronted with the proof, insisted that she knew nothing about it. Even though the police found strong evidence connecting her to the incident, she continued to maintain that she had no memory of it.

Because of the conflicting testimonies, her attorney asked that she have a thorough medical examination. She was proven normal in all respects, except that the blood-sugar test revealed she had hypoglycemia.

The discussion in the *British Journal of Psychiatry* states:

> ... Doctors and others who have been associated with diabetes mellitus or with insulin shock therapy are aware of the serious derangement of cerebration and eventual loss of consciousness caused by hypoglycemia.... On the day of the accident the patient had fasted for about 21 hours and had taken a carbohydrate meal 3 to 4 hours before the accident; our investigations showed that this course of events would be highly likely to have brought her to the time of the accident in a severely hypoglycemic state, though not necessarily incapable of driving a car

automatically. Within an hour prior to the accident she had taken a glass of sherry, and it is known that a small quantity of alcohol is capable of enhancing the effects of hypoglycemia.... The patient was charged with causing death by dangerous driving. ... At the trial only medical evidence was called by counsel for the defense.... Counsel was able to show the court that the defendant suffered from reactive hypoglycemia, and that at the time of the accident she was probably hypoglycemic to the extent that her state of consciousness was altered. The patient pleaded 'guilty' as a matter of fact, even though she had no memory whatsoever of the accident. A nominal fine was imposed, with a driving disqualification on the grounds of present ill health.[1]

A doctor's wife who is an upstanding citizen, an average American man who has never before committed a crime—killing because of hypoglycemia? Yes, it happens. Low blood sugar is no respecter of persons, age or status. A baby can be born with it; a person can die because of it. Sometimes it takes an extreme trauma or stress to "trigger" it so that symptoms become manifest. Often no overt signs are apparent until it is too late. No one is immune from its potential ravages, and it often causes distress early in life.

Jill's symptoms, for example, started with migraine headaches at age four. She had all kinds of food allergies and became irritable and nauseous when she ate the wrong things. By the time she was thirteen she was suicidal. Her parents took her to many doctors, experiencing the same kind of medical merry-go-round that I had gone through.

When Jill married at nineteen, her husband watched her on the merry-go-round. Before she was thirty-five, she had visited *forty-three* doctors and had two children – a girl, six, and a boy, eight. Her daughter, Lori, was having similar problems and, in addition, had a terrible disposition. Much of the time she cried or threw tantrums. Her pediatrician excused the daily headaches by saying, "It's a stage she's going through," or "It's growing pains." Bobby, her son, was hyperactive and had both learning and behavior problems.

Partially because of the stress of two growing, unhealthy children in her life, Jill's headaches became worse, her tiredness turned into exhaustion, and she became suicidal again. One day she happened to hear Dr. Carlton Fredericks talking about hypoglycemia on the radio. He discussed the symptoms listed in his book, *Low Blood Sugar and You*,[2] and suggested that listeners with any of those symptoms read his book.

Jill bought it, and discovered that, out of the sixty-five symptoms listed, she had sixty-four! She was sure that she had found the solution to her problems. Her own physician, however, discounted the idea, but at her insistence, he agreed to give her a three-hour glucose tolerance test. It proved to be negative.

Thus she began her search for a doctor more knowledgeable in this area to give her a six-hour test, as Dr. Fredericks suggested. She finally found him in Dr. Norman East. After administering the six-hour test, he confirmed that she had hypoglycemia and placed her on the proper diet.

"I made a spectacular recovery, and I owe my life to that man," Jill says. "In a short time I felt better than I had ever felt in my whole life. Imagine! I was thirty-five years old and had never known what it was like to experience happiness or *not* to be sick. I'll always have to stick to the diet, but it's worth it to be able to function like a normal human being. If I eat a teaspoon of sugar I get a migraine. It's just not worth it, so I stay away from the stuff. I wouldn't go back to where I was the first thirty-five years of my life for anything."

After Jill's low blood sugar was discovered, she began to suspect that her children might have it, too. Their pediatrician refused to accept that possibility, however, and would not test them. So Jill took them to her doctor. Sure enough, Bobby's test was positive, and he was put on the hypoglycemia diet. Rather than give the glucose tolerance test to a child as young as Lori, the doctor diagnosed her empirically. That is, he assumed from her symptoms and history that she was suffering from hypoglycemia and put her on the diet.

The changes in the children were as radical as those in their mother. In a short time, what had been a pain-racked, highly disturbed family became a physically-well, happy one.

Often, however, metabolic problems surface in a victim's early teens. Jeff, at fourteen, was an all-round good kid, doing well in school, liked by his teachers and peers, was never in any trouble, and had a paper route at which he worked diligently.

Suddenly, things began to change. He became indifferent about his school work, didn't care whether or not his papers were delivered, and often started to pick fights with those around him. His five brothers and sisters were afraid to go near him, and the students at school stayed as far from him as they could. He hit his mother so much that she had to wear long-sleeved blouses to hide the black and blue marks.

The family doctor suggested a psychiatrist. When the psychiatrist was unable to help, his parents took him to another. Jeff only continued to be violent so they went to another. This therapist suggested they take him to Napa, a state hospital for the mentally ill.

Wanting to know more about the institution before they committed their son, they decided to look it over. After a Sunday afternoon drive to Napa, their decision was easy.

"I wouldn't put my worst enemy in there!" Jeff's father fumed.

A few days later Jeff threw a knife at one of his brothers and struck him in the leg. The police were called, and he was taken to juvenile hall. Eventually, the authorities sent him to Napa. He was there eleven months when the doctors said he was well enough to come home. Away from the stresses of school, his paper route, and large family, he hadn't been violent once. They thought he had calmed down for good.

Nevertheless, as he and his parents were on their way home in the car one Sunday afternoon, he declared without warning: "You've taken eleven months out of my life, and I'm going to get you for it." Evidently, the resentment against his parents had been building.

His abusive behavior resumed, and his mother again had to wear her long-sleeved blouses. They knew he needed more help. The psychiatrist they found to handle his case this time was "low blood sugar conscious." Immediately recognizing the symptoms of hypoglycemia, he ordered a glucose tolerance test for Jeff. It was positive, and he advised them to take him back to the family doctor for treatment.

I later talked to that doctor to learn why he hadn't kept Jeff in his care.

"Adults will need psychological help, along with the diet, to get out of the negative behavioral patterns they've established over the years," he explained. "A child as young as Jeff will probably get better on just the diet, so I didn't feel he needed my help."

But unfortunately, Jeff's doctor gave him the diet

for diabetics, which only worsened his condition. They took him back to the psychiatrist, armed with the diet that his doctor had given him. When he was placed on the proper diet, his family watched him improve. It wasn't long before his behavior became normal. He has since finished high school, graduated from college and become a fine, productive member of society. Few people know that he was once a patient in a mental hospital.

As I think of Jeff's experience, I wonder how many children with similar problems have become life-long residents of such institutions. How many adults with unsuspected and undiagnosed hypoglycemia are permanent inhabitants of such hospitals? The number could be staggering. Some authorities believe that as many as eighty percent of those in mental hospitals have nothing wrong with them but hypoglycemia.

Similarly, the stresses related to attending college can also trigger hypoglycemia. Bob, a student at the University of California, became impatient, tense, and irritable. His mother, herself a hypoglycemic, suggested that he get a glucose tolerance test. But the father of Bob's best friend was a doctor who didn't acknowledge hypoglycemia. He convinced Bob that his mother's problems were "all in her head."

Believing this, Bob went to see a psychiatrist about his difficulties. After months of counseling, he still felt terrible. In spite of his problems, he was a brilliant student and graduated with high honors. When he married, his physical troubles continued, and his wife insisted that he see a doctor. He was given a glucose tolerance test, and his blood sugar went down to thirty.

"Hardly anyone could deny that he was a severe hypoglycemic," his mother told me, "but he still doesn't believe that means anything, and he won't do anything about it. He grew up with the idea that 'there's nothing wrong with me; it's all in my head.' He's uptight and feels awful, but refuses to believe that his low blood sugar has anything to do with it."

Frequently, symptoms of hypoglycemia do not manifest themselves until midlife. John, for example, was in his late thirties when his difficulties started. His problems weren't severe, but they were of definite concern to him and his family. A successful professional man, he had his own insurance firm and enjoyed a good business on the Monterey Peninsula where he lived. His life was interesting and satisfying. He had a lovely wife who was a good mother to their two children, a gourmet cook and a wonderful hostess when they entertained in their beautiful mountain-top home. It would seem that John had everything to be happy about, yet he found himself crying at the least little thing. Besides the crying, he was sleepy most of the time, and couldn't seem to control his anger.

"I thought I was going crazy," John told me. "Since I was into hypnosis at the time, I tried to hypnotize myself. That didn't work, so I had a doctor friend hypnotize me. That didn't help either. Then I went to a hypnosis seminar where one of the speakers was Dr. Harold W. Harper, a physician of North Hollywood, California. Dr. Harper passed out a form to each person in the audience and asked us to check off the things that applied to us. He said that all sixty-five things listed could be symptoms of hypoglycemia. He advised that if we had hypoglycemia, we needed a change of diet, not hypnosis.

"He explained that if you have a metabolic disorder, which hypoglycemia is, then no amount of hypnosis or counseling can correct that difficulty."

John later went to Dr. Harper for a glucose tolerance

test which showed that he definitely had a sugar problem. His fasting blood sugar level was 100, which is in the normal range. But then it went up to 214 and down to 58. His doctor said that any fall of more than fifty points should be considered drastic and indicative of hypoglycemia.

Dr. Harper gave him a copy of the hypoglycemia diet, and instructed him to be sure to eat every two or three hours. Soon the anger, sleepiness, and crying stopped.

"It was clear that the way my body was handling sugar was the problem," John told me. "Whenever I cheated on the diet, it would take a couple of days for my body to get rid of the sugar and get back to normal. For about three months I couldn't handle sugars at all, but now as long as I don't cheat too much, my body can take a little."

John is rather unique in being able to tolerate small amounts of sugar because most sufferers must follow the diet all their lives. It seems that one simply cannot live a normal life as a hypoglycemic. The problem of low blood sugar is no different than being blind or deaf or a paraphlegic. It is a physical condition that must be dealt with daily.

All these stories demonstrate that hypoglycemia affects everybody differently. Our metabolisms are as varied as our fingerprints. While some of us exhibit similar symptoms, hardly any two people have the same set. For many, the distress may be slight or only mild. But for me, it was hell on earth for almost five years.

Physical symptoms associated with hypoglycemia not only wreaked havoc with my body, they greatly distressed my soul. I began to realize that thousands of individuals like myself also are languishing in needless guilt and frustration over seemingly spiritual problems when the trouble is in reality physical. They cry out in despair, wondering whether God has abandoned them. For these desperate victims and the loved ones who suffer with them, there is hope. Ample evidence is available to uncover the masquerade and end the feelings of guilt that torment them.

5/Where Are You, God?

"Help me . . . Help me . . . God, please help me."

I can't begin to describe the futility behind that call—a cry that was constantly on my lips when I was suffering so severely. When the children were in school, I would lie on my bed for hours and scream those words over and over again. Even the hypoglycemia diet hadn't helped, and tranquilizers were ineffective. The pressures of my responsibilities—especially the raising of three young teenagers—were so intense that everything within me cried out, "I can't stand it much longer!"

No one seemed to understand—or care—at that time. Even going to church didn't seem to help. Every morning I managed to feed the children and get them off to school, throw a coat over my short housecoat and go to church. Once there, I sat in the last pew and cried. I couldn't pray because all I could think of was myself. I knew I wasn't a hypochondriac, as my sister often accused. Their illnesses are imaginative; mine was metabolic. But that didn't lessen my desperation.

Such a sense of hopelessness can greatly hinder the spiritual life of hypoglycemia victims. Struggling with feelings of fatigue and depression, they begin to question their relationship to God and endure much guilt as a result of their emotional swings.

Pastor Jim Roberts, for example, was as despondent as many of his parishioners who came for counseling. In bewilderment, he sought advice from a trusted friend, the pastor of a nearby church.

"I'm beginning to think I should get out of the ministry," Jim shook his head solemnly. "Sometimes I feel like a hypocrite. People are coming to me for guidance because they're full of fear and anger and guilt. Sometimes they're so depressed they don't even want to live. I wonder what they'd think if they knew that I feel the same way most of the time. I've been so depressed lately...."

On the surface, Jim always seemed happy and cheerful, his messages often exhorting his congregation to "rejoice in the Lord." Indeed, he was a minister who believed what he preached. Yet something was wrong. In frustration, he sought help.

"Maybe your problem isn't a spiritual one," his friend offered. "Maybe you should be talking to a doctor instead of me."

"I've been to my doctor, and he's given me all kinds of tests. He says everything is okay, but I *know* something's wrong. If it isn't physical, and it isn't spiritual, then I must be going crazy. Sometimes I really feel as if I'm losing my mind!" Jim confessed.

Gazing sadly out the window of his friend's study, he recalled a frightening experience. "The other day I was driving down the freeway, and all of a sudden I hardly knew where I was. I was completely disoriented. In a daze, I pulled off to the side. I must have blacked out because it was dark when I came to. And that's not the first time it's happened! I just can't go on this way. Maybe I should just quit the ministry ...."

"Do you really think you'd feel better if you did?"

Jim stared pensively into space and sighed. "I don't know. I just don't know. I really don't want to, but I can't think of anything else to do."

"Why don't you see another doctor. Maybe he can find something...."

The "other" doctor was Dr. Walter Johnson, a neurologist, who put Jim in the hospital a few days later. He and another specialist, an endocrinologist, conducted a series of tests. Jim was lying quietly in his bed when Dr. Johnson stepped into the room with the results.

Coming right to the point, he said, "Jim, we don't know yet exactly what's going on in your head. There's some dysfunction of the brain that would indicate a tumor. We need to do some more tests, and possibly operate. But before we do, we want you to go home and get things in order."

Jim sighed with relief when Dr. Johnson mentioned a possible solution, but the words "put things in order" jolted him.

"You mean," Jim stumbled over the words, "you mean this could be fatal?"

"The prognosis is not good. You could live and have some loss of mental capacity – more loss of brain function than you have now. You probably won't be able to pastor, because it could be a very fast-growing tumor. My advice is, go home and get ready for whatever might happen."

Disturbing thoughts raced through Jim's mind as he drove home.

How do I get ready for the worst? What'll Pat do? She's never worked except at home since we've been married. Will the kids be able to go to college? Could those doctors be wrong? Should I see another specialist? If so, who?

He heard the answer to the last question as soon as he walked into the house. Dr. Johnson had called, wanting Jim to see Dr. William Young, an ear, nose, and throat specialist. Jim made an appointment for the following week, arranging to have his records transferred.

Dr. Young was sitting at his desk when a nurse escorted Jim into his private office. He looked up and smiled.

"I've just been going over your file," he said as he rose to shake hands. "You've certainly had lots of tests, haven't you?"

"Yes, I have," Jim agreed as he took the chair offered to him.

"Tell me, what's bothering you now?"

"We-l-l-l, my headaches have been almost constant for several months. I've had them all my life."

"Are you having any vision problems?"

"Yes. I often see double, even triple."

"What else?"

"I don't have any energy, even though I get plenty of rest. I sleep like a baby for seven or eight hours, but I'm more exhausted when I wake up than when I went to bed. Also, I have this funny feeling inside. I don't know how to describe it—it's sort of like my motor's running all the time, and it won't stop."

Dr. Young smiled again and asked, "Would 'inside trembling' describe it properly?"

"Yes, that's it!"

"It's not too common, but it happens." Dr. Young's voice was soft and matter-of-fact, as though he were trying to put Jim at ease.

"I thought I was going crazy, that I was imagining it. I've never heard of anything like that. I've been so depressed at times I didn't want to live."

"Did you have these problems as a child?"

"Just the headaches, and a lot of sore throats."

"Any trouble learning?"

"No, but a lot of trouble trying to sit still in school."

"You'd probably be called hyperkenetic today," Dr. Young observed, smiling.

"Jim, all these things are symptoms of hypoglycemia or low blood sugar — from your hyperactivity when you were a boy to what's going on today. I think that's causing your problems, but there's only one way to find out — a glucose tolerance test. I'm going to order one for you, and don't let anybody do anything else until we get the results. I've seen many cases like yours where a brain tumor is suspected, and all that was wrong was hypoglycemia."

The possibilities of a tumor were dispelled when the test results came in. Dr. Young explained that Jim did have this metabolic problem, which was correctible only by a change in his nutritional pattern. He gave him a copy of the hypoglycemia diet and instructed him to follow it faithfully.

After three days on the diet, Jim experienced his first relief from depression in months. Within a few weeks, the headaches had completely disappeared, his energy level was back to normal, and he was bustling about, performing his duties as usual.

God's people are not insulated from depression. The psalmist David suffered from it saying,

> Why are you cast down, O my soul,
>   and why are you disquieted within me?
> I say to God, my rock:
>   "Why hast thou forgotten me?
> Why go I mourning
>   because of the oppression of the enemy?"[1]

> Vindicate me, O God, and defend my cause
>   against an ungodly people;
> From deceitful and unjust men
>   Deliver me!
> For thou art the God in whom I take refuge.... [2]

David was despondent because his people were being oppressed by the enemy—deceitful and unjust men. It was clear that his problems came from outside sources, not from an upset metabolism. He *knew* that God was his answer, and he called upon Him.

The solution to a hypoglycemic's troubles, however, is often not as simple as David's. Many times, God's people try to turn to Him, but find themselves unable because their physical condition hinders them. As prayer and Bible reading become ineffective, their doubts and fears often replace faith, leading to guilt and self-condemnation. They exist in a mental prison which can be unlocked only by proper diagnosis and treatment, which includes proper nutrition and self-discipline.

Although God's power can heal hypoglycemia, it usually takes some personal effort to stop putting the wrong things into our bodies, to eat food the way God made it, and to get rid of many other errors of living.

Another pastor who is familiar with hypoglycemia stresses the importance of good nutrition. "The first thing I do when I'm counseling people who are depressed," he says, "is to ask them about their physical condition and what they eat. I believe much depression comes from eating the wrong food, especially sugar. The body affects the spirit's sense of well-being, and the spirit tremendously influences the body. If a person feels bad, he will have trouble fighting off depression, or praying, or doing anything spiritual. His sagging strength will keep him from using the energy of God."

The answer to depression caused by hypoglycemia does not lie in more prayer, more Bible reading, or more involvement in church. It rests in the discipline of our bodies.

The Apostle Paul taught that our bodies are temples of the Holy Spirit, and that we should glorify God in them. Yet, millions of Christians are treating their stomachs as though they were garbage cans, not temples. Coffee and cigarettes are the biggest addictions in America. Many people drink gallons of coffee, much of it loaded with sugar. White flour pastas, jello salads (chief ingredient, sugar), pastries, and other sugary desserts find their way to family reunions and church potlucks.

But nature simply does not excuse ignorance of the laws of nutrition. The more we ignore these rules, the more sickness there will be. Researchers have shown that degenerative diseases have increased in America in proportion to the consumption of sugar.

Considerable care is needed to avoid foods containing artificial products. In the "good ole days," foods were sold in their natural state. Not too long ago, H. J. Heinz was about the only food manufacturer around, and the consumer ate most foods the way they were grown. Now, almost everything is processed, preserved, added to or changed in some way. Today the best we can do is train ourselves to eat as little unnatural food as possible.

Such self-discipline in nutrition was one thing my young friend Becky had to learn. As a little girl, she was always hungry, especially for sweets. And if they weren't given to her, she'd find a way to get some.

"I've been sick most of my life," Becky confided. "When I was little, I was known as the weak one of the family. My leg muscles were always so sore that

I couldn't run and play like the other kids. My mother said I just had growing pains, but she couldn't explain why I was always so hungry for sweets."

When I met Becky, she was twenty-six, married and had two children. She had known for about a year that she had hypoglycemia. As usual my first question was, "How did you find out about your low blood sugar?"

"My doctor suspected it when I started hyperventilating. After my second baby was born, I was having all sorts of problems. I was afraid of everything, and was really depressed. Harold and I thought my problem was mental or spiritual, so I started going to a counselor a couple of times a week. But when I started hyperventilating, Harold insisted I go to the doctor."

The physician suspected hypoglycemia and ordered a glucose tolerance test. It confirmed his suspicions, and he immediately put her on the proper diet. She soon began to feel better.

But Becky admitted that it took more than a change in diet to make her feel better. "I've had to really discipline myself. The doctor said I needed lots of rest, and that's been hard. Sometimes I've had to make myself stop working so I wouldn't get exhausted. I always felt guilty about not getting my work done. I used to not eat breakfast and would skip lunch, if I were busy. That's really been hard to change."

Feelings of guilt can cut deeply when our spiritual perceptions become clouded and distorted. Rosine, my pillar-of-the-church-type friend, suffered severely in this way. Conscience-stricken because she had no energy and couldn't go to work, she slipped into depression. She wondered how she could be so close to God, who is all peace and joy, and still be unhappy. She didn't feel like going to church, couldn't concentrate long enough to read her Bible, and could hardly pray. As feelings of guilt mounted, Rosine sank deeper into despair.

"Sometimes I feel like God's abandoned me," she wept. "Deep down I know that's not true, but it *feels* like He doesn't really care."

As her tiredness and depression grew worse, I tried to tell her about hypoglycemia. She couldn't believe that a physical problem could be the root of her troubles and refused to see a doctor.

"Rosine, I *know* hypoglycemia can cause problems like yours!" I insisted. "I had them myself, and I've met many others with the same symptoms, and they had it too. Please go to a doctor!"

Because I pleaded and coaxed for more than a year, she finally weakened.

"Maybe you're right. I'm going to pray about it and ask God to show me in some way that I need to have that test."

Twice in the next week, Rosine tuned into her favorite television program—the 700 Club—and heard two doctors discussing low blood sugar. Both advised the glucose tolerance test for people who were suffering from tiredness and depression without apparent cause. Taking this as a sign from God, she made an appointment for the test. The results confirmed my suspicion. She did have hypoglycemia. After a week on the diet, she felt much better.

"I can hardly believe it," she rejoiced. "I'm glad you kept after me. I never realized that food could make such a difference in the way I feel spiritually, or in the way that I see God."

In addition to depression, many other symptoms of a metabolic disturbance can affect one's spiritual well-being. Anxiety, negativism, anger, paranoia, agitation, fear, crying spells, restlessness, antagonism, compulsions, temper tantrums, apathy, irritability, worrying, thoughts of suicide, belligerence, anti-social behavior, vague sense of dread, emotional instability, and lapse of moral conduct—all take their toll. Under such burdens, one certainly has little or no inclination to pray. And many people, such as Rosine, even feel abandoned by God.

Because most of these feelings are unreservedly categorized as sin, victims of hypoglycemia become laden with guilt. Worry, fear, crying, and a vague sense of dread may indicate a lack of trust in God, but it is not always the case. Negative emotions produced by a physical condition can affect our ability to reach out to Him or to sense His presence in times of trouble without indicating faithlessness. Where emotional problems have a spiritual basis, sin may indeed be the underlying cause. In such cases, one can overcome his difficulties through recommitment to God and by obedience to His Word. Nevertheless, physical causes for emotional stress cannot be overlooked.

In his book, *Nutrition and Your Mind*, Dr. George Watson observes:

> Among the first things that happen when the blood sugar is too low and sufficient glucose is not available to the brain is loss of normal emotional control. This can take many forms, from simple nervousness, unexplained weeping and depression, all the way to violent impulses, the immediate urge to smash something—anything.[3]

Perhaps out of ignorance sometimes, spiritual advisors tend to overlook the possible effects of the physical on the spiritual. Frequently, their counsel is to pray, read the Bible, and get more involved in the church. However, if a physical condition *is* the underlying cause of a spiritual problem, this approach will be ineffective. Better advice, in some cases, might be: "See a doctor, have a glucose tolerance test, and eat properly."

This dilemma, however, is complicated by the differences of opinion which surround the medical profession. The memory of my painful ride on the medical merry-go-round prompted me to examine closely both sides of the controversy over the diagnosis and treatment of this national health menace. I wanted to learn *why* hypoglycemia was a stepchild of medicine — ignored, put down, even denied. If it did not exist, then how could it be wrecking the lives of millions, especially without their knowledge?

Part 2
Understanding Hypoglycemia

6/Stepchild of Medicine

"Good morning, KMST."
"Could you tell me if you have any talk shows?" I inquired.
"Yes, we have one at noon called 'Midday.' Doug Moore is the host. Would you like to talk to him?"
"Yes, please."
A click on the other end signaled that my line was placed on hold. Seconds later a pleasant voice broke the silence.
"Hello. This is Doug Moore."
"Doug, my name is Jinny Zack. I live in Carmel. There's something I'd like to talk about on your show. It's called hypoglycemia. Know anything about it?"
"I never heard the word until recently. My secretary just found out she has it. She's been going to doctors for months trying to find out what's wrong with her."
"There are a lot of people who have it and don't know it. That's why I want to make as many people as I can aware that it could be causing their problems. A lot of doctors don't seem to be paying any attention to it, and it can be pretty serious."
"Sounds as if it would make an interesting show," he injected. "The trouble is that I don't have any openings for three months. But ... the man who's supposed to be on today might not show up. I forgot to call and remind him. It's already ten o'clock and we go on at noon."
"What if I come out to the station and sit around

till then? If he shows up, fine. If he doesn't, I can go on."

"That sounds like a good idea." His voice brightened. "Do you have a book or something on hypoglycemia that I can look at?"

"Lots of them. I'll bring some with me."

The enormity of what I was doing didn't hit me until I was speeding along the freeway toward the television studio. *What am I thinking of? I hardly ever watch television, and now I'm going to be on it. What about make-up? Are these clothes all right? What am I going to say?*

I began to ponder the message I wanted to get across. Would people accuse me of practicing medicine without a license? I certainly didn't want to give that impression. I only hoped to make them aware of a condition that is often overlooked. Because I had searched for help for so long, I wanted to spare others that kind of a pursuit. If one person listening to me would be helped, it would be worth the effort.

Pushing my misgivings into the background, I concentrated on the task at hand. By the time I arrived at the studio, I was ready. Waiting for Doug to appear, I leafed through the pamphlet from the Hypoglycemia Foundation[1] which I had brought with me. Seated at his desk moments later, I handed him the booklet and answered a few questions as he perused it. Suddenly, he glanced at his watch.

"It's quarter to twelve! I guess he's not coming, so let's go."

He led me to the sound stage where the studio crew was busily getting things ready. We took our places on the set and talked about the program as the microphones were being adjusted. In no time we were on the air.

Looking back on that day, I wonder at my nerve! Perhaps it was the same determination that had brought me through those many months of depression and turmoil. However, appearing on that television program was easy and fun. I felt good about sharing my knowledge, which I felt *had* to be voiced because so many people were suffering needlessly from a condition that was being largely ignored or underestimated.

What made me think that I, single-handedly, had to try to change this state of affairs? The answer was simple: I had read enough to know that few doctors were doing much to correct the situation. My research had shown me, not only that hypoglycemia was a stepchild of medicine, but *why* it was.

Low blood sugar, sometimes considered life-threatening, was recognized in 1924 by Dr. Seale Harris who was doing research on diabetes at the University of Alabama. Just a few years before that, one of the greatest discoveries of the century had been made—insulin—used in the treatment of diabetes.

Some think that diabetes is almost as old as history; others believe that it only came into existence with the widespread use of sugar. When I was a child, it was always referred to as sugar diabetes.

For years, nothing had been able to stop the ravages of this dread disease, which afflicted millions of persons whose systems were not producing enough insulin. Finally, Dr. Frederick Banting, a Canadian orthopedic surgeon, joined the search for a way to manufacture insulin because of his interest in a neighbor's child. He was distressed that there was nothing to alleviate her suffering.

Charles A. Best, a physiologist, joined Dr. Banting in his experiments, and they soon discovered what others had long been seeking—the way to produce insulin outside the body.

With this discovery, almost every diabetic in the world began to get regular injections of insulin. Until they were able to regulate their dosages, however, many diabetic patients received too much, and went into insulin shock. But when Dr. Harris noticed that many of his *non*-diabetic patients also were going into insulin shock, he began to wonder if their systems could be creating excess insulin.

Much testing proved that he was right! He had discovered hyperinsulinism (too much insulin) which later became equated with hypoglycemia (too little blood sugar). This over-abundance of insulin was causing the blood sugar to fall too suddenly or too low.

Many have asked me, "If low blood sugar is your problem, then you just eat more sugar, right?" This is a dangerously wrong assumption.

Hypoglycemia is too little *blood* sugar, which is entirely different from the sugar we eat. The kind that we put into our coffee and tea, the stuff that goes into candies, pastries, ice cream, and most processed foods—that sugar, and related malnutrition, is one of the problems and not a solution. Processed or refined carbohydrates, especially sugar, white flour and white rice, stimulate the pancreas to produce too much insulin too quickly. These and three other substances— caffeine, alcohol and nicotine—drastically upset the sugar metabolism and can cause insulin shock. A quick drop in the blood sugar level, or a fall to a particularly low level, causes a plunge in the effectiveness of our body mechanism.

The efficiency of the fuel we put into our bodies governs the way our systems work. Despite what the television commercials say and what Madison Avenue and the sugar industry would have us believe, sugar is *not* an efficient fuel and does not provide lasting energy. A 1970 study showed that the ability of the body to oxidize glucose (the process by which we obtain energy) is reduced if the source of that glucose is refined sugar.[2]

Since all the foods we eat are processed or metabolized in our bodies to provide nourishment for our cells, the sugar we consume changes so fast that it gives a spurt of energy. However, for individuals with upset sugar metabolisms — hypoglycemia or diabetes — this is where the trouble starts. A hypoglycemic's pancreas is over-stimulated and liberates too much insulin, a hormone that then causes the body to burn up sugar. A diabetic's pancreas, however, is not stimulated enough, and so does not produce enough insulin.

The central nervous system, the body unit most vulnerable to hypoglycemia because it cannot store blood sugar, has to rely on constant feeding. If the supply is too low, the system is starved. That's why hypoglycemics must carry food with them all the time. They need a protein snack every two or three hours because protein metabolizes slowly, but eventually makes for more energy.

The brain, the key part of the nervous system, is the first to feel a deficiency when you have hypoglycemia. Sometimes it hardly works. To illustrate, picture yourself standing in front of the bread shelves in a grocery store, frustrated because you can't decide which brand to buy. Imagine standing in front of rows of canned soups thinking, *I must be going crazy. I can't make up my mind.* See yourself going wild trying to figure out what kind of meat to choose from a vast sea of packages. If you have experienced any of these anxieties, you could have hypoglycemia.

One of my first symptoms, along with tiredness and

being unable to make decisions, was an inability to subtract figures in our checkbook. Because all at once I couldn't keep the numbers straight, Jay had to take over that job. Although, at the time, I presumed I was distressed by nothing more than overwork, I now realize that my starved brain couldn't concentrate, make decisions, or do simple math. Distress over the inability to do simple tasks clouded my mind and triggered frequent emotional outbursts.

But hypoglycemia has more than emotional symptoms; many physical problems can be traced to it. Diseases of the pancreas, the liver, thyroid, and gonads (ovaries and testicles) usually develop in later life, but the adrenal glands can be affected even before birth. A baby who starts life with colic, diaper rash, respiratory disorders, and various reactions to formulas (most of which have some sort of sugar) may have an adrenal problem. This often manifests itself in hypoglycemia, which is usually overlooked.

A baby can be born with deficient glands if the mother's are defective. When pregnant, a hypoglycemic draws from the healthy glands of the fetus. For this reason, many hypoglycemic women feel better during pregnancy than at any other time. But this also explains why the baby can be born with faulty glands.

The adrenals provide the hormones that a pregnant woman needs most. And since her own body isn't supplying them, she steals them from the fetus. This problem need not be, because adrenal cortical extract (ACE) can correct it easily. Although not a potent material, its continuous and regular use will build up the adrenals so that they will eventually work properly on their own.

Dr. John W. Tintera, an endocrinologist, thoroughly researched ACE early in his notable career, and used it on thousands of his hypoglycemic patients. A specialist in glandular disease, Dr. Tintera devoted most of his life to the study and treatment of hypoglycemia. As the guiding light of the Adrenal Metabolic Research Society of the Hypoglycemia Foundation for many years, he discovered that, since the adrenal glands regulate sugar metabolism, they are critical in the treatment of hypoglycemia.

Early in my research, I met a young nurse who had worked with Dr. Tintera. She was hypoglycemic and convinced of the value of ACE. Knowing that hypoglycemia can be hereditary, she wanted to do everything possible to keep her children from having it. During both of her pregnancies, she had ACE injections regularly. Both of her babies were born with healthy adrenals and normal blood sugar levels.

The proper use of ACE could prevent other serious problems. Some doctors believe that infant hypoglycemia may account for many mysterious crib deaths (Sudden Infant Death Syndrome), and if their mothers had received ACE, their babies would have been normal.

The July 1971 issue of *Postgraduate Medicine* had some interesting things to say about hypoglycemia in infants:

> Hypoglycemia also occurs in infants of subnormal weight, including those who are premature, small for gestation, and the smaller of twins.... Hypoglycemia also has been noted in unusually large infants.... The incidence of hypoglycemia is increased in infants born to women who have had toxemia of pregnancy.... Children of RH incompatible mothers have a 4 percent incidence of hypoglycemia....
> Hypoglycemia has been considered a consequence

of hypothermia (low temperature).... The injudicious use of salicylates (aspirin), and alcohol sponging for fevers have produced hypoglycemia.[3]

Certain signs of low blood sugar can be detected in older children. One is the same tired-all-the-time-for-no-reason feelings that adults often have. Frequent sore throats, headaches, nightmares, convulsions, vomiting for no apparent reason, forgetfulness, car sickness and learning and/or behavior problems are others.

One reputable pediatrician, when asked about low blood sugar in children, declared that there is no such thing as pediatric hypoglycemia. However, a computerized survey of medical journal articles, which covers the period from 1965 to 1982, lists more than two hundred reviews of hypoglycemia in infants and children. Several of them are listed in *Appendix A*.

For example, the *Medical Journal of Australia* stated in its July 4, 1970 issue:

> The newborn child at the time of birth usually has the same blood glucose level as its mother.... Particularly at risk are those of low birth weight.... This occurs frequently in the smaller of twins and infants of toxemic mothers, suggesting that inadequate nutrition of the fetus in utero predisposes to the postnatal development of hypoglycemia. Unusually large newborn infants seem also at risk.... There is strong pathological evidence that the neonatal brain will suffer extensive neuronal loss if subjected to hypoglycemia for more than a few hours.... With early detection and treatment in maternity hospitals, a significant reduction in the numbers of brain-damaged children in a community should be obtainable.[4]

Fortunately, the diagnosis of hypoglycemia in adults is not so nebulous. Dr. Harris named the following symptoms in his early writings:

Weakness	Irritability	Low blood pressure
Nervousness	Constipation	Narcolepsy
Headaches	Drowsiness	Mental deficiency
Dizziness	Epilepsy	Cold hands and feet
Fatigue	Depression	Worrying unnecessarily[5]

Since that time, myriads of symptoms have appeared in the literature. It is interesting to note that some are the exact opposite of others, such as high and low blood pressure, fatigue and hyperactivity, underweight and obesity, sweating and cold. Dr. Currier explains this by saying that faulty diet or stress can cause our body chemistry to fluctuate from one extreme to the other.

It should also be noted that a person need have only one or two of these symptoms to have hypoglycemia. By the same token, everyone need not have the same signs, because our metabolisms are as different as snowflakes. If a person has an undiagnosed illness, however, and one or several of these difficulties are present, hypoglycemia should be suspected and tests given.

The following warning signals, gleaned from medical literature, are listed according to the frequency in which they are mentioned:

- Anxieties
- Tremors (inside or out)
- Heart palpitations (fast beating heart)
- Mental confusion
- Hunger
- Lack of concentration
- Blurred vision
- Insomnia
- Crying spells
- Convulsions
- Sweating
- Alcoholism
- Allergies
- Sleepiness, even narcolepsy
- Personality changes, schizophrenia or mental illness
- Anti-social behavior
- Fears and phobias
- Fainting or blackouts

Poor memory
Loss of sex drive (frigidity in women, impotence in men)
Muscle pain, twitching or spasms
Cold, clammy skin or cold sweats
Obesity
Psychosis or neurosis
Restlessness
Addiction (or craving for) sugar, caffeine, soft drinks, cigarettes, drugs
Incoordination
Suicidal tendencies
Arthritic pains or arthritis
Ulcers (peptic or duodenal)
Digestive disorders
Staggering (loss of muscle control)
Numbness
Asthma
Gasping for breath
Nightmares or night terrors
Speech difficulties
Juvenile delinquency
Sleep disturbances, especially sleeping few hours, waking up, and can't go back to sleep
Poor appetite
Gas on the stomach
Stomach burning
Sudden increase or decrease in weight
Angina
Dimness of vision
Double vision
Disorientation
Unconsciousness

For less frequently mentioned symptoms, see *Appendix B*.

Just as with symptoms, there can be numerous causes of hypoglycemia. According to the lay press, the chief underlying cause is stress—physical, mental, emotional or dietary. Physical pressure can surface as infection, pain, over-exertion, child-bearing, burns, fractures, surgery, or an overdose or wrong use of drugs. Mental strain can develop while cramming for exams or with other heavy intellectual work. Emotional tension may include the death of a family member (even that of a pet), divorce, loss of a job, financial problems, or any of numerous other daily tensions. Dietary stress can result from over-use of the wrong foods, particularly processed carbohydrates (including *any* kind of sugar, white flour, and white rice), caffeine and alcohol.

According to most of the lay literature, *any* prolonged stress plays a significant role in blood sugar control. Too much tension can result in the exhaustion of overworked glands. Dr. Paavo Airola, one of America's foremost nutritionists, says that anything that causes stress and strain on the system can contribute to the development of hypoglycemia.[6]

Because so many sources listed sugar as a culprit, I decided to do further research in that area. Of thirty-three articles on hypoglycemia appearing in national magazines between 1930 and 1981, only half a dozen examined causes. Each contended that excessive use of sugar in the average American diet is the primary cause of hypoglycemia. Some discussed other contributing factors such as too much refined starches, coffee, tobacco, alcohol, heredity, and the daily stresses and emotional pressures of American life. Today, book stores and the book sections of health food establishment are teeming with volumes condemning sugar.

The "sugar-is-bad-for-you" theme started to thread its way through the periodicals in 1958 when *Time* reported on a study done by Harvard's School of Dental Medicine Biochemist James H. Shaw. Dr. Shaw's investigation showed that all kinds of sugar promoted tooth decay, and he recommended that we cut down on sugar consumption. He also reported that his work had stopped because the Sugar Research Foundation which had funded his study had withdrawn its support.[7]

In 1973 *Good Housekeeping* published an article called "Sugar, Can It Be a Health Hazard?" The writer's answer to the question can be found in the article's subtitle—a list of the problems he believes sugar causes: *Obesity, Dental Decay, Heart Diseases, Hypoglycemia, Diabetes, Arteriosclerosis, Infections.*[8]

Another writer asked: "What's wrong with sugar? Why do respected nutritionists say that sugar is helping to make the American diet a national disaster? For one thing, we eat too much of it. A high sugar diet can strain the body metabolism in harmful ways, but sugar has been advertised as 'energy food' for so long that people think they have to have it, which isn't true."[9]

"Sugar as we know it," commented another, "is as much an industrial product as gasoline."[10]

"Warning: the surgeon general may someday determine that sugar consumption is dangerous to your health," cautioned another article.[11]

Most authorities agree whole-heartedly with these comments. Others go a step further, suggesting that a skull and crossbones be put on each box, bag, or bowl of sugar.

One writer asked this question: "Which contains a greater percentage of sugar – Heinz tomato ketchup or Sealtest chocolate ice-cream, Wishbone Russian dressing or Coca Cola, Coffee Mate non-dairy creamer or a bar of Hershey's chocolate?" Here are the surprising answers:

Heinz ketchup, 29%	Sealtest ice cream, 21%
Wishbone dressing, 30%	Coca Cola, 9%
Coffee Mate, 65%	Hershey bar, 51%[12]

Amazingly, even iodized salt contains sugar. And for pets, there's plenty of it in most brands of their foods.

Merely the titles of these articles speak volumes:

"Sugar, Enemy of Good Nutrition"[13]
"Sugar, Coronaries Linked"[14]
"Sugar: Those Silent Sugar Diseases"[15]

"Killer on the Breakfast Table"[16]
"Sugar, Villain in Disguise"[17]
"Sugar—How Sweet It Is—and Isn't"[18]
"Sugar—No. 1 Murderer"[19]

Only one of the above articles came from a health publication. The rest are from nationally-known, general-interest magazines.

Why is sugar harmful? Processed sugar is a crystallized form of sucrose. It is refined and adulterated (prepared for sale by replacing more valuable with less valuable or inert ingredients in whole or in part). Our bodies were created to eat whole, natural food the way God made it, not the way man changes it. Sugar cane or sugar beets fresh from the field are full of nutrition. Processing them, however, removes between eighty-nine and ninety-eight percent of the minerals, vitamins, trace elements, enzymes, fatty acids and proteins.

Processed sugar is a pure chemical. In no way can it be considered a food, which is defined as "a nutritive material taken for growth, work or repair and for maintaining the vital processes."

Dr. K. W. Donsbach has compiled a list of hidden sugars in some of the common foods we eat. The amounts are amazing:[20]

"Hidden Sugars" in Foods

Food Item	Size of Portion	Approximate sugar content in teaspoonfuls of granulated sugar
	Beverages	
Cola drinks	1 (6 oz. bottle or glass)	3½
Cordials	1 (¾ oz. glass)	1½

90 / *Sugar Isn't Always Sweet*

Ginger ale	6 oz.	5
Hi-ball	1 (6 oz. glass)	2½
Orange-ade	1 (8 oz. glass)	5
Root beer	1 (10 oz. bottle)	4½
Seven-Up	1 (6 oz. bottle or glass)	3¾
Soda pop	1 (8 oz. bottle)	5
Sweet cider	1 cup	6
Whiskey sour	1 (3 oz. glass)	1½

Cakes and Cookies

Angel food cake	1 (4 oz. piece)	7
Apple sauce cake	1 (4 oz. piece)	5½
Banana cake	1 (2 oz. piece)	2
Cheese cake	1 (4 oz. piece)	2
Chocolate cake (plain)	1 (4 oz. piece)	6
Chocolate cake (iced)	1 (4 oz. piece)	10
Coffee cake	1 (4 oz. piece)	4½
Cup cake (iced)	1	6
Fruit cake	1 (4 oz. piece)	5
Jelly-roll	1 (2 oz. piece)	2½
Orange cake	1 (4 oz. piece)	4
Pound cake	1 (4 oz. piece)	5
Sponge cake	1 (1 oz. piece)	2
Strawberry shortcake	1 serving	4
Brownies (unfrosted)	1 (¾ oz.)	3
Chocolate cookies	1	1½
Fig Newtons	1	5
Ginger snaps	1	3
Macaroons	1	6
Nut cookies	1	1½
Oatmeal cookies	1	2
Sugar Cookies	1	1½
Chocolate eclair	1	7
Cream puff	1	2
Donut (plain)	1	3
Donut (glazed)	1	6
Sweet roll	1 (4 oz. piece)	4½

Candies

Average chocolate milk bar (example: Hershey bar)	1 (1½ oz.)	2½
Chewing gum	1 stick	½
Chocolate cream	1 piece	2
Butterscotch chew	1 piece	1
Chocolate mints	1 piece	2

Fudge	1 oz. square	4½
Gum drop	1	2
Hard candy	4 oz.	20
Lifesavers	1	⅓
Peanut brittle	1 oz.	3½

Canned Fruits and Juices

Canned apricots	4 halves & 1 tbsp. syrup	3½
Canned fruit juices (sweetened)	½ cup	2
Canned peaches	2 halves & 1 tbsp. syrup	3½
Fruit salad	½ cup	3½
Fruit syrup	2 tbsp.	2½
Stewed fruits	½ cup	2

Dairy Products

Ice cream	⅓ pt. (3½ oz.)	3½
Ice cream bar	1	1-7 depending on size
Ice cream cone	1	3½
Ice cream soda	1	5
Ice cream sundae	1	7
Malted milk shake	1 (10 oz. glass)	5

Jams and Jellies

Apple butter	1 tbsp.	1
Jelly	1 tbsp.	4-6
Orange marmalade	1 tbsp.	4-6
Peach butter	1 tbsp.	1
Strawberry jam	1 tbsp.	4

Desserts, Miscellaneous

Apple cobbler	½ cup	3
Blueberry cobbler	½ cup	3
Custard	½ cup	2
French pastry	1 (4 oz. piece)	5
Jello	½ cup	4½

In the same way as natural sugar, wheat, after the whole grain is refined, becomes a nutrition-deficient white powder. More than ninety percent of the good is taken out, leaving mostly starch. From two to four percent of the grain's nutritional value is returned in the form of synthetic vitamins. White flour losses through milling are incredible as shown in the following chart:[21]

Nutrient	Whole Wheat Flour	(70% extraction) White Flour
Protein	11.8	10.8
Oil	2.26	1.14
Carbohydrate (as starch)	66.0	73.0
Thiamine	3.6	0.7
Riboflavin	1.87	0.61
Niacin	52.45	9.35
Pyridoxine	5.7	0.11
Pantothenic Acid	11.1	7.1
Biotin	0.08	0.024
Folic Acid	0.37	0.2
Iron	34.0	11.0
Sodium	33.0	21.5
Potassium	3405.	975.
Calcium	319.	160.
Magnesium	1250.	206.
Copper	6.3	2.0
Zinc	34.8	10.8
Phosphorous	3450.	915.
Choline	375.	470.

Four items – thiamin, riboflavin, niacin and iron – are put back in the flour, and they call it *enriched* white bread!

Eating sugar and white flour presents several other difficulties. First, devitalized products satisfy hunger but leave no room for nutrition. Second, continuous use of processed food can eventually lead to nutritional deficiencies and cause starvation. And third, since our bodies are designed to handle *only* natural, whole foods, continual consumption of processed ones places a great strain on many glands and organs, causing them to function improperly. Wholesome food, under the supervision of our adrenal glands, keeps the amount of glucose and oxygen in the blood in balance. When sugar and other denatured food escape processing in our bodies, they are absorbed directly into the blood stream. Thus, the precise balance is destroyed, and the body is in crisis.

Professor John Cawte, director of the Intercultural Mental Health Center in Sydney, Australia, says, "The excessive use of sugar is more of a health threat in the Western world than abuse of alcohol, tobacco, or narcotics."

The amount of sugar eaten in the United States soars each year. In 1900 the average American ate seven pounds a year. Today, it is 120. In one study it was found that some teenagers eat as much as 400 pounds a year.[22]

But in spite of the growing opposition to sugar, one strong and stubborn voice is heard in its favor – that of Dr. Frederick J. Stare who has earned the title, "Harvard's Sugar-Pushing Nutritionist." Dr. Stare is the founder of the Department of Nutrition at Harvard's School of Public Health. On the building in which he works is a gold-lettered marble plaque which reads:

> The Harvard School of Public Health gratefully acknowledges the generosity of General Foods Corporation in making possible the Nutrition Research laboratories within this building.

General Foods is one of the largest users of sugar in America. Dr. Stare and a few associates continue to maintain that it is good for consumption and that more should be used.

Yet, many authorities believe that if sugar were a new drug and had to be approved by the Food and Drug Administration, it wouldn't be allowed on the market. Because the FDA considers sugar a food and not an additive, which it is technically, tests have never been made for its safety.

It is interesting to surmise what would happen if sugar were analyzed as other additives are—and found harmful. Removed from the products we consume daily, sugar might well be proven to be the culprit causing multiple mental, emotional, spiritual and physical disorders.

7/Millions Suffer Needlessly

Depression, crying spells, sleeplessness, confusion, irritability, personality changes, suicidal tendencies, loss of sex drive, psychosis or neurosis, antisocial behavior—these are only a few of the signs that indicate you could be among the millions of Americans who suffer from hypoglycemia.

As we have seen, much disagreement exists over whether such symptoms really indicate low blood sugar. What a physician believes about hypoglycemia, says Dr. Currier, depends greatly upon what journal or textbook he studied while in medical school. He believes that a physician's emotional involvement with the subject also is responsible for his bias. The hotter his emotions, the more biased he seems to be.

Doctors often feel threatened when their patients suggest what might be wrong with them, Dr. Currier confesses. "Many physicians think that if they don't know anything about a subject, the knowledge doesn't exist. They get quite upset if patients give opinions or mention something they have read.

"In forty years of practice, I have come to realize that we physicians have a fragile ego. Some of us are afraid to admit that we don't know something; we believe that our patients will think less of us if we don't have all the answers. Many times we are undeservedly put on a pedestal by our patients.

"How often have you heard the saying that the doctor 'tries to play God'? I have a lovely, bright, intelligent eighty-four-year-old patient with a good sense of humor, whom I see for her nutritional-metabolic-preventive medicine problems. Because she was having some mild heart difficulties recently, I suggested that she see a cardiologist. A few weeks ago she told me that this other doctor is one of the kind I just described. She described her last encounter with him succinctly saying, 'But for the grace of God, there goes God.' "

A rapidly increasing number of physicians, however, are recognizing that mental and emotional difficulties can be caused by a physical condition such as low blood sugar. Understandably, the medical profession is embroiled in controversy over this issue because many symptoms of hypoglycemia are characteristic of mental and emotional problems. Therefore, most doctors discount the possibility of low blood sugar in their evaluations and too often send their troubled patients to psychiatrists for help.

Following Dr. Seale Harris' early research of low blood sugar, most mental and emotional disturbances were listed as symptoms of this disorder. With the advance of psychiatry, however, those ills were given over to the professionals in that field. And wrongly so, for most of these therapists did not consider the possible physical reasons for mental and emotional stress. In the process, hypoglycemia has become one of the most neglected and untreated conditions of today, and millions of its victims are suffering needlessly.

Curious about why this controversy exists, I returned to Lane Medical Library at Stanford University, planning to search the professional journals for a few articles on the subject. From a computer print-out of 2,300 listings of articles gleaned from 2,800 of the world's bio-medical journals, I chose approximately 200. My study of those works helped me to understand the reasons for the controversy surrounding the condition.

In 1933, nine years after his original work on low blood sugar, Dr. Harris disclosed the findings of his research on six thousand hypoglycemic patients. Because of his studies, he came to the conclusion that the underlying cause of the disorder is a diet deficient in vitamins which contains an over-abundance of processed sugar. He wrote:

> Sugar-saturated vitamin-starved Americans, that is, those who live largely on white flour bread, white potatoes, white rice, sugar-saturated coffee, sugar-laden desserts, with candy and soft drinks between meals, would seem to be prone to become victims of pancreatic disorders, including hyperinsulinism (later equated with hypoglycemia).[1]

Between then and 1950 only one of twenty-three articles surveyed discussed causes. It was written by Dr. Sidney A. Portis who supported Dr. Harris' viewpoint.[2] From then to the present only two references to the correlation between wrong diet and hypoglycemia were noted. Nutrition was *never* mentioned in the medical journals. Why?

One reason soon became clear. In the early 1970s, not one medical school in the United States taught a course on nutrition. A few offered an hour's class on the subject, but that was optional. Consequently, doctors learned little or nothing about the importance of proper food to our well-being unless they happened to study it on their own after going into practice.

Fortunately, today, that picture has changed. Twelve of the 129 medical schools in the country have a required nutrition course.[3] These are listed in *Appendix C.* Approximately half have elective courses.

Only two dozen of all the journal articles discussed the *causes* of hypoglycemia which are given in the list below. Those in italics were mentioned most often.

Ethonol (alcohol)

Other drugs, especially aspirin, oxidase inhibitors (enzymes which promote oxidation), or sulfonylreas (agents which lower blood sugar, such as Orinase, Diabonese, Dymelor, and Tolinase)

Early diabetes

Alimentary (after stomach surgery)

Ideopathic (of no known origin)

Iatrogenic (doctor-induced, the consequence of treatment of diabetics with excessive amounts of insulin or sulfonylreas)

Excessive strenuous exercise

Renal glycosuria (sugar in urine)

Liver disease

Hypopituitarianism (low thyroid function)

Adrenal deficiency

Thyroid deficiency

Inborn errors of metabolism (born with deficient glandular function)

Vomiting in childhood,
convulsions in children

Scleroderma (skin disease)

Lactation (producing milk)

Malabsorption of food

Cirrhosis of the liver

Hormone deficiency

Pregnancy

Enzyme deficiency

Tumors of the pancreas (listed everywhere as a rare cause)

Poor dietary intake

One glaring fact is evident here. *Absolutely no mention was made of the dangerous effects of certain foods.* Even though "poor dietary intake" was discussed two or three times in two hundred articles, no allusion was made as to what that might mean.

Alcohol-induced hypoglycemia is on almost every list of causes. Many nutrition-minded doctors believe that *all* alcoholics have hypoglycemia. Dr. Currier explains it this way: "The alcoholic's liver is placed under a great strain because it has to work hard to break down alcohol without the needed materials to carry on its work. It begins to store excess fat and becomes less able to do its real job of saving glycogen, the storage form of blood sugar. The victim cannot keep his blood sugar level up."

Alcoholism has been classified by some as a disease. If so, the source of the ailment has not been pinpointed. It might be that the basic cause is an upset sugar metabolism, in which case proper diet, *along with* the Alcoholics Anonymous program, would be a most effective means of treatment.

When I consider everything a hypoglycemic must give up until his sugar metabolism is back to normal, I recall the story of one doctor's dilemma. Since his hypoglycemic patients had to cut out sweets, white flour, coffee, and alcohol, he hated to make them cut out cigarettes, too, because they might start cutting out paper dolls!

Tumors of the pancreas are listed, both in the lay press and the medical journals, as a *rare* cause of low blood sugar. Yet the National Institute of Health, an agency of the United States government, has been studying hypoglycemics with pancreatic tumors for several years. Hearing about this program through my congressman, I called the Institute in Washington, D.C. and was referred to their Institute of Arthritis, Metabolism, and Digestive Diseases, the Diabetes Branch. I asked the doctor with whom I spoke why the agency was doing so much research into a rare condition. "Our studies," he explained, "are related to the secretion of insulin, which is vital in diabetes. Diabetes, not hypoglycemia, is our chief concern."

"Why don't you study other kinds of hypoglycemia?"

"Because they're not life threatening," he responded. "They might be disconcerting, but not life threatening."

It would seem to me that any disorder which leads to such severe depression that it results in suicide, or to alcoholism which can cause all kinds of violent deaths, *must* be considered life threatening.

I pressed the doctor about the fact that hypoglycemia often leads to diabetes. "Don't you think that if hypoglycemia were discovered and controlled in early life, there wouldn't be so much maturity-onset diabetes?"

"There is no evidence," he replied, "that hypoglycemia will have any effect on future diabetes."

Contrary to his opinion, however, there is *much* evidence in medical literature that low blood sugar can be, and often is, a pre-diabetic state. In 1933 Seale Harris wrote that clinical evidence suggests that the patient with hypoglycemia is a potential diabetic. Therefore, early dietary management may prevent diabetes.[4] Similarly, in 1956 Drs. Holbrooke S. Seltzer, Stefan S. Fajans, and Jerome W. Conn were also proponents of this theory.[5] In 1962 Dr. Robert Gittler wrote:

> Rapid advances have been made by investigators who have concentrated predominantly in the study of spontaneous hypoglycemia. Significant contributions have been achieved in ... the recognition of spontaneous hypoglycemia as an early manifestation of diabetes.[6]

In March 1970 Drs. Albert S. Finestone and Michael G. Wohl advised, "It has been known for some time that spontaneous hypoglycemia may occur as an early manifestation of diabetes mellitus."[7] In 1977 Dr. Paul T. Chandler reported that "the principal causes (of hypoglycemia) are alimentary, functional ... or early diabetes mellitus."[8]

Apparently, many physicians failed to see these studies. If a cure were found for hypoglycemia, perhaps the number three killer in the United States—diabetes—could be lessened or eliminated. If hypoglycemia were discovered early in a person's life and kept under control, much maturity-onset diabetes could be avoided.

Further research also showed that physicians authoring books for the lay press and those writing in the medical journals are poles apart. Therefore, I turned to the textbooks, hoping to find some agreement.

Since hypoglycemia is an internal problem, I surveyed texts on internal medicine and on endocrinology—the study of the endocrine glands which are critically important to the sugar metabolism—and one book on medical diagnosis and treatment. Unfortunately, in checking the sections on hypoglycemia in each book, I found four different viewpoints.

Among the dozens of causes for hypoglycemia given in these textbooks are endocrine deficiencies, liver disorders, substrate deficiency, gastric surgery, counter-regulatory deficiency, sensitivity to insulin, neonatal, hypoglycemia in children of diabetic mothers, erythroblastosis fetalis, leucine sensitivity, alcohol-induced, pancreatic tumors, starvation, malabsorption of food, hereditary fructose intolerance, galactosemia, and ideopathic (of no known origin.)

The textbooks had only three similarities. All included alcohol-induced hypoglycemia and ideopathic hypoglycemia, and omitted any reference to food except malabsorption. However, because more and more doctors are adding poor nutrition to the list of causes, perhaps someday ideopathic hypoglycemia will be identified as a nutrition problem.

Reflecting on the information gleaned from textbooks, I wondered what text was used at Stanford to teach about hypoglycemia. A call to the Endocrinology Department revealed the astonishing answer.

"Doctor, I'm doing some research on hypoglycemia and would like to ask you a question," I began.

"Fine, go ahead."

"What textbook do you use to teach the subject?"

"Oh, we don't use a textbook. As soon as they're printed, they're obsolete."

"Then where do the students get their information?"

"From our lectures and from the medical journals."

"You mean what they learn depends upon which journal or author they read, and what they remember of your lectures?"

"Yes, that's essentially correct."

Unwittingly, that doctor explained another reason for the confusion about hypoglycemia. The disagreement, not only between the lay press and the medical profession, but among the doctors themselves, is astonishing. The line seems to be drawn between the nutritionists and nutrition-minded doctors and drug-oriented physicians.

Dr. Robert C. Atkins, author of *Diet Revolution* and other books on nutrition-medicine gives insight as to why these factions can't get together. Since organized medicine is supported by the economically powerful drug industry, he maintains, many doctors don't want to lose that support by replacing drug therapy with proper nutrition. He writes:

> There can be no question that the widespread acceptance of nutrition medicine as the first line of defense against illness (rather than drugs which are now the first defense) would be seen as a major threat to the drug industry.[9]

The drug-industry-organized-medicine-combination of which Dr. Atkins speaks has another powerful ally in this struggle, the food manufacturers. There's practically no canned, packaged, or processed food on the market that doesn't contain at least one, and often two or three, forms of sugar. It is being used, not only as a sweetener, but as an additive, a filler, a texturizer, and a preservative.

With the indiscriminate use of not only sugar but other additives and chemicals, the food industry may be unintentionally contributing to many diseases, including hypoglycemia. Low blood sugar is the basic cause of many diseases and can be controlled by proper nutrition as opposed to drugs. But acknowledgement of this fact by drug oriented factions would mean admitting that drugs are the second line of defense against disease and that proper nutrition is more important.

Dr. Atkins also contends:

> It would not be surprising if your own doctor, who has been subtly influenced through his medical education, the repeated denial of low blood sugar's existence in his journal reading, as well as the same media we have all been exposed to, whitewashing our national diet, may be reluctant to investigate the diagnosis of hypoglycemia or even to recommend the good nutritional practices which correct it. Further, now that malpractice suits have become his most distressing problem, he may even be fearful to take a position opposed by his leadership.[10]

Much evidence supports the claim that hypoglycemia can be a cause of mental and emotional disorders. Dr. Harris noted many emotional and mental signs of hypoglycemia, as did other doctors of that era. His work on hyperinsulinism was highly recognized by the American Medical Association when he received its Distinguished Service Medal.[11]

In 1927 Dr. P. J. Cammidge of London wrote that indications of hypoglycemia are:

> A vague feeling of uneasiness.... The patient becomes faint, anxious, excited and emotionally unstable, vertigo (dizziness) may be complained of ... delirium, disorientation, delusions, confusion, and bradycardia (slowness of heart beat) may occur.[12]

In 1935 Dr. James Greenwood Jr. reported that hypoglycemia is a cause of mental symptoms, and recorded many cases from the Psychotic Department of Philadelphia General Hospital to back his claim. A higher percentage of diagnosed low blood sugar was noted in psychotic patients in the general medical and surgical wards. He discovered two distinct characteristics among the mental patients: one was a psychosis which took almost any form—sometimes violent; the other was psychotic manifestations which alternated with periods of normal behavior.[13]

From that time until 1950, many articles appeared in journals concerning the mental and emotional states caused by hypoglycemia. Some of their titles are thought-provoking:

Hypoglycemia as a Cause of Mental Symptoms	1935[14]
Personality Changes in Hypoglycemia	1936[15]
Psychological Problems in Hypoglycemia	1943[16]
Psychiatric Aspects of Spontaneous Hypoglycemia	1948[17]
Hyperinsulinism, a Factor in Neuroses	1949[18]
Life Situations, Emotions and Hyperinsulinism	1950[19]

Besides the above, other evidence indicated that emotional and mental disturbances were associated with hypoglycemia. Dr. Joseph Wilder, a prolific writer of his time, wrote:

> (In many hypoglycemics) there is a general slowing down of mental functions. Consciousness becomes clouded, thinking is hampered by weakness of concentration, and the emotional state is characterized by depression or timidity. Pronounced irritability and the tendency to opposition may assert themselves.
> ...The purely psychological changes include serious

> impairment of thinking power to the point of complete blocking, inability to fix attention, and loss of power for coordinating and abstracting intellectual work.... (This) may lead to disorientation and mental haziness...extreme degrees of "dawdling," apathy and indifference...and negativism.... Hallucinations and other "psychoasthenic" phenomena may occur in this condition.[20]

More evidence was gathered from Drs. David Adlersberg and Henry Dolger. They reported:

> There is a wide variety of mental changes displayed during hypoglycemia, from mild anxiety or exhilaration to severe psychotic states.... (There may be) lack of will power and inability to make simple decisions.[21]

Most of Dr. Harris' original reports and those of the physicians just quoted were written before World War II. During that war, America's doctors toiled in the service or in the understaffed offices and hospitals of our nation, having little time for reading, writing, or research.

By the time America returned to normal, psychiatry had become popular, and mental and emotional illnesses were relegated to the psychiatrists' couches. For many years they stayed there, and hardly anyone remembered that these problems could also have a physical basis.

In the late 1960s, however, some physicians became aware that poor nutrition, an upset metabolism, and hypoglycemia are possible causes of mental illness. Dr. Harry M. Salzer wrote a long dissertation entitled "Relative Hypoglycemia as a Cause of Neuropsychiatric Illness," in which he declared that hypoglycemia can mimic any mental disorder. He listed the major symptoms of hypoglycemia that he thought should be considered from a psychiatric viewpoint: depression, insomnia, anxiety, irritability, lack of concentration, crying spells, phobias, forgetfulness, confusion, unsocial or antisocial behavior, psychosis, and suicidal tendencies.[22]

A trend toward this thinking began to develop. In several parts of the world, psychiatrists were discovering that many of their schizophrenic patients were suffering from hypoglycemia. This led them to a thorough study of nutrition and metabolism. Some eventually founded an organization called the Academy of Orthomolecular Psychiatry in London in 1971.

In 1975 a group of doctors who were interested in nutrition also assembled in California. They established the Orthomolecular Medical Society whose membership includes not only psychiatrists, but other nutrition-oriented doctors. Membership now includes physicians in other states and Canada. The society's purpose is defined in their literature:

> Orthomolecular medicine is the treatment or prevention of illness by providing the optimum molecular environment for the mind and body. Illness results from an imbalance within the body resulting from improper diet, exposure to toxins – heavy metals, allergens, etc. – faulty genetic inheritance or prolonged emotional or physical stress and the resultant excesses or deficiencies which constitute the imbalance.
>
> Orthomolecular physicians are concerned with the total functioning of the body and consequently are concerned with metabolic tests which indicate imbalances within the body. Among the tests used are the hair analysis which gives an analysis of minerals and heavy metals, the glucose tolerance test which analyzes blood sugar levels, urine samples, and

various blood tests designed to locate vitamin, mineral or hormonal excesses or deficiencies.[23]

While many doctors joined this organization to treat the mental and physical together, more and more medical journals included less and less about the mental and emotional effects of hypoglycemia. Nonetheless, *some* physicians wrote about them in books and magazines. A survey of twenty-five books and twenty-nine magazine articles dealing with hypoglycemia showed that only thirty-four discuss symptoms. Each of them declares that mental and emotional problems can be a sign of hypoglycemia.

Adding to the confusion, a statement appeared in 1973 denying that certain physical and emotional symptoms are indications of low blood sugar. It was issued by three well-known medical organizations: the American Medical Association, the American Diabetes Association, and the Endocrine Society. It appeared in four medical journals, even though such articles are usually published only in the journal of the organization issuing them. This statement, therefore, received four times as much coverage in the medical world as it would have ordinarily. It was also released in the public media.

The statement declared, among other things, that:

> There is no good evidence that hypoglycemia causes depression, nervous breakdowns, chronic fatigue, childhood behavior problems, allergies, alcoholism, juvenile delinquency, drug addiction, and inadequate sexual performance.[24]

Since I had seen all of the above listed as symptoms in various places, I decided to go over this list item by item with one question in mind: "Were the authors of the books and magazine articles wrong, or were the issuers of the statement in error?"

The conflict of viewpoints is summarized in the following chart:

The statement says hypoglycemia does not cause	Number of books, magazines and medical journals which says it does
Depression	28
Chronic fatigue	29
Nervous breakdown	37
Alcoholism	7
Juvenile delinquency	7
Childhood behavior problems	14
Drug addiction	5
Allergies	7
Inadequate sexual performance	4

The nine conditions which the statement says hypoglycemia does *not* cause are declared symptomatic at least *138* times in the literature I surveyed. Depression, fatigue, and nervousness are among the signs most-often mentioned, not only in the lay press, but in the medical journals. "Chronic" fatigue is included only a few times, but is usually classified as "extreme," "excessive," or "undue." Nervousness is the symptom most frequently listed in the journals between 1924 and 1973, the year of the statement. Along with insomnia, mental confusion, negativism, phobias, rapid heartbeats, and profuse perspiration, nervousness certainly contributes to emotional problems.

Juvenile delinquency, the statement also asserts, is not caused by hypoglycemia. Yet the following articles from medical journals illustrate the contrary:

Problem children...are frequently found to have a tendency to hypoglycemia...Certain types of crimes and offenses, characterized by increased aggressivity, lack of self-control, loss of moral inhibitions, impairment of judgment, are apt to be committed in hypoglycemia. They are accompanied...by certain psychological features: irritability, a rebellious attitude against the representatives of any kind of authority, a general tendency to negativism, apparent indifference in the matters concerning ethical and social conventions. The patient yields easily to any kind of asocial impulses, stopping not even at self-destruction...Certain features are considered characteristic of certain types of criminals, particularly the defective delinquent. But the difference is that these features are promptly and completely reversible in hypoglycemia, while they are deeply ingrained in the criminal. This remarkable reversibility...affords perhaps an unequalled field for experimental research in normal, pathological and criminal psychology.[25]

Another example of the practical importance of the study of hypoglycemia is *crime*. This may sound surprising unless we remember how many psychological features we have found in hypoglycemia which *might* result in crimes and transgressions....We compiled a whole list of cases from the literature and our own experience of crimes committed in the state of spontaneous or induced hypoglycemia, like disorderly conduct, particularly resistance against the police, assault and battery, attempted homicide and suicide, cruelty against children, like sticking a pin into a baby's eye, etc., matrimonial cruelty, various sexual perversions and aggressions, false fire alarm, embezzlement, petty larceny, willful destruction of property, arson, slander and the frequent violation of traffic regulations.[26]

These observations were written long before the 1973 AMA statement and seem to describe precisely what is happening in America today.

Drug addiction is another "nonsymptom" mentioned in the statement, although several authorities, in at least sixty-three articles, indicate that drug addiction *is* caused by hypoglycemia. A relationship between drug abuse and low blood sugar definitely exists; the question is—which came first, the drug dependency or the hypoglycemia?

Until precise and controlled studies are done to measure and detect all the metabolic changes which co-exist in these conditions, hypoglycemia will remain a foggy area of medical knowledge. Unless more objective evidence is obtained, no one can state with certainty that they are not caused by hypoglycemia.

In fact, some studies have already been undertaken. Such experts as Dr. John Yudkin of England, Drs. Emil Cheraskin and E. M. Abrahamson, and nutrition expert-author-lecturer J. I. Rodale have charged that excessive sugar consumption and its often accompanying hypoglycemia is the major contributing factor in the increased rate of crime, drug addiction and alcoholism in the United States.

Inadequate sexual performance, the last item on the statement's list of problems that hypoglycemia does not cause, is another debatable issue. Although Dr. Robert Atkins includes a lack of sex drive in women and impotence in men in his list of symptoms, perhaps research in this area has been limited. Low blood sugar in itself may not have caused my inadequate sexual performance, but constant fatigue, irritability, crying, negativism, mental confusion, fears and phobias certainly contributed to my lack of interest in sex.

Thus, the statement that "there is no good evidence"

for certain hypoglycemia-induced problems is doubtful. Who is to say whether the evidence gathered in this great nutrition-medicine debate is good or bad? At any rate, testing for hypoglycemia should be a routine part of physical examinations of persons with these symptoms, particularly if no other cause has been found. Clearly, until the marriage of nutrition and medicine occurs, millions will continue to suffer needlessly, for only then will the evidence be good enough to satisfy both factions.

8/Discovering the Enemy

by Wilbur D. Currier, M.D.

Hypoglycemia is not a disease; it is the result of an extremely complex perversion of body chemistry. Nearly every function of the body is involved in blood sugar metabolism. For this reason, I prefer the term dysglycemia to hypoglycemia or even to diabetes, because both are the result of perverted blood sugar metabolism. For the sake of continuity, however, I will stay with "hypoglycemia."

Perhaps no word in the medical dictionary is more fraught with misunderstanding than this one. Almost two dozen books have been written on the subject; hundreds of articles have appeared in magazines, medical journals and textbooks. Because few of the authors agree on the cause, diagnosis or treatment of this condition, it is little wonder that doctors are uncertain as to what hypoglycemia is, or even if it exists.

Since the discovery of low blood sugar, the generally accepted method of diagnosis has been the glucose tolerance test (GTT). However, that is only one of the many means I use in detecting and finding the cause of hypoglycemia.

The patient who suffers from the symptoms of hypoglycemia has to be examined in a basic, thorough manner. Each side of the testing triangle—history, symptoms, and laboratory tests—is valuable in the diagnosis. The history is most important, because it gives up to eighty-five percent of all the information necessary to establish a diagnosis and treatment.

In my office we use three detailed history forms. Two are filled out by the patient and one by the doctor. The first gives information which only the patient can report—symptoms which he is experiencing. These often are numerous and involve almost any system of the body, including the mind and emotions.

I pay particular attention to one item in my patient's history, which is the question, "Is or was there any diabetes in your family?" If the answer is yes and the patient is having symptoms of hypoglycemia, I can be reasonably certain that his blood sugar level will be abnormal.

The second form used is a computerized diet survey on which the patient gives a detailed report of the kinds of food he usually eats, and how often. This extensive review is sent to a laboratory where it is run through a computer. The results show whether the patient has significant deficiencies or excesses in his diet.

Since environmental factors are vital in these days of polluted air, chemicalized foods and other harmful substances, we look for unsafe levels of minerals in the system. Hair analysis is a useful tool in this process, for it is a reasonably accurate evaluation of the minerals entering our tissue cells. Because our bodies are composed of approximately sixty trillion microscopic cells, we are either sick or healthy, depending upon the chemistry inside those cells. Hair, being living tissue, reflects the mineral excesses or deficiencies of intracellular chemistry.

The Heidelberg Gastric Analysis is a method we use routinely to determine whether a patient is absorbing his food by secreting adequate amounts of hydrochloric acid. A deficiency of that substance means that the foods he eats, particularly the protein, will not be broken down properly and be prepared for absorption into the blood stream. This is called *malabsorption*, which is similar to not eating properly.

Malabsorption is one of the most common reasons for erratic blood sugar levels and is what causes so many symptoms to surface. When food is not being absorbed into the system and transformed into energy, problems will arise.

The hydrochloric acid in the stomach is the chief mechanism by which animal protein is broken down into its component amino acids. These amino acids can then be absorbed through the bowels into the bloodstream and are built back into human protein. Also the hydrochloric acid, in breaking down the protein into its component amino acids, permits these amino acids to grab onto, or to chelete the minerals that are ingested into our system. If this mechanism is not functioning properly, the patient is getting little benefit from what he is ingesting into his stomach. This, of course, could lead to relative starvation.

Knowing the level of hydrochloric acid in a patient's stomach is vital. It is reasonable to assume that if the person has a deficiency in hydrochloric acid (achlorhydria), he is also lacking in other digestive enzyme substances in the mouth, stomach and bowel. Hence, a complete well-rounded digestive enzyme preparation containing hydrochloric acid must be prescribed.

On the other hand, some individuals can have excessive amounts, and it would be wrong for them to take this preparation. Doing so could lead to a gastric or duodenal ulcer problem.

Frequently hormone studies are necessary to uncover deficiencies in that area. Tests must be run to determine thyroid function, adrenal cortical gland operation, estrogen, female or male hormone activity and, at times, insulin and pituitary function. It is my belief that if one endocrine gland is sick, the others will be too. I usually find that the pancreas isn't the only gland to be affected by hypoglycemia.

Allergies must also be considered in the diagnosis. We can suspect these chiefly from the history and, at times, physical findings. They can mimic almost any kind of disease.

However, one of the most misunderstood and misread clinical tests in medical science is the glucose tolerance test which, in my office, is performed routinely on all new patients. Since it has been estimated that from fifty to eighty percent of all Americans have low blood sugar *to one degree or another,* GTTs should be given to all patients.

Sugar diabetes can be detected by the analysis of the sugar content in one drop of blood. If this were true of hypoglycemia, the diagnosis would be much simpler. But the doctor who is looking for hypoglycemia needs to know, not only the amount of sugar in the blood at one given time, but how the body reacts to a specific amount of sugar solution during several hours of testing.

Although some physicians prefer to give a six-hour test, I have found that significant changes seldom occur in the sugar level after the fifth hour. Therefore, I give five-hour tests unless the patient requests six.

The glucose tolerance test reveals how toxic 100

grams of glucose is to an individual. That amount will be poisonous to everyone because the human body has no enzyme system to metabolize refined carbohydrates. However, some people have a sound body chemistry which protects them against the ravages of that much refined carbohydrate.

I pay little attention to the readings on the GTT. My primary concern is how the patients feel while they are taking the test. This gives me a good indication of their fundamental body chemistry reserve which protects them against the effects of sugar, white flour and junk foods.

To prepare for a glucose tolerance test, patients must have no food after dinner the evening before, though I permit them to have small amounts of water. They should not smoke or take any drugs during this time, unless they are absolutely essential. The test usually starts at 7:30 or 8 in the morning.

When the patients first arrive at the laboratory, a sample of blood is taken from a vein in the arm. After this, they drink 100 grams of glucose. In a half hour a second sample of blood is taken; the third, in another half hour. The first half hour sample is vital, but one which many doctors omit. Another sample is then obtained every hour for four more hours. The patient remains in the office in a restful state without smoking or eating throughout the test.

The only aspect of the test upon which most physicians agree is what a normal one should be. The consensus is that the fasting blood sugar should be between 70 and 120, that the level should go up considerably (there is little agreement as to how high), and that it should then fall back to what it was in the fasting state and remain at that level for the rest of the test.

The accepted "normal" test looks like this:

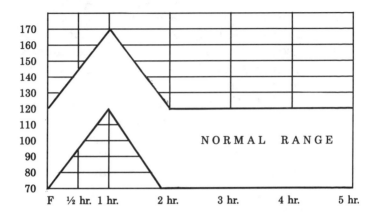

Any significant variation from this represents an abnormal blood sugar chemistry, or dysglycemia.

Opinions about the GTT, as well as other aspects of blood sugar chemistry, are chaotic. Some doctors take half-hour samples throughout the test; others take them every fifteen minutes for a certain period. There are many arbitrary variations.

Length of the test is also disputed. Dr. Seale Harris realized that a three-hour test was useless. "Diabetes can be diagnosed in three hours," he said. However, it has become clear that, although hypoglycemia *can* be detected in that length of time, it is not safe to assume that it always *will* be. In many cases, it doesn't appear until after the fourth or fifth hour. Unfortunately, many doctors give three-hour tests for hypoglycemia which are essentially invalid and prove nothing.

Once the GTT has been administered, pandemonium about its interpretation begins. There are almost as many ways to interpret the tests as there are doctors who interpret it. In one of his earliest papers, Dr. Harris reported that patients who have symptoms and show readings of 70 milligrams of sugar per centimeter of blood have hypoglycemia.

Today most physicians insist that the blood sugar reading must drop to a certain level before they will diagnose hypoglycemia. The favorite number seems to be 50 (milligrams of sugar per centimeter of blood), although I have seen 60, 55, 45 and even 40 mentioned in medical journals and textbooks. These writings indicate that any figure used is arbitrary. In other words, they are based on opinion, not fact.

In 1952 Dr. Allen O. Whipple issued a history-making report. He had made a life-time study of islet cell tumors of the pancreas, one cause of hypoglycemia. Dr. Whipple set forth a thesis which has since been generally accepted by the medical world as "Whipple's Triad." This precept was to be used *only* in the detection of organic hypoglycemia—that caused by a pancreatic tumor. None of the three points, either collectively or individually, were meant to be used in diagnosing regular hypoglycemia, known as reactive, functional or spontaneous, which is the topic of this book.

Whipple's Triad for diagnosing *only* organic hypoglycemia requires that:

1. The patient must develop the hypoglycemic attack during the fasting state.
2. Twenty-four hour fasting blood sugar levels below 50 mgm %, or below 50 mgm % at the beginning of or during an attack must be demonstrated.
3. Immediate recovery from the attack on administration of intravenous glucose.[1]

Since doctors mistakenly use the second criteria—blood sugar level *must* go below 50—to diagnose functional hypoglycemia, it is no wonder that many cases of spontaneous hypoglycemia have been undiagnosed, or misdiagnosed, and ultimately mistreated. Drs. Allan L. Drash, Max G. Ellenberg and Holbrooke S. Seltzer, for example, wrote in 1972 that hypoglycemia is often missed.

> When this happens, the patient can suffer irreparable brain damage particularly if the initial attacks occur during infancy or childhood or if he is an adult with a significant metabolic alteration.[2]

No reference is made in medical literature before Dr. Whipple's treatise of 1952 which asserted the sugar level had to reach 50 or any other number before a diagnosis could be made. Medical science is aware that man is an individual and has unique metabolic traits. Each person's metabolism is different and what is "normal" for one is abnormal for another, and vice versa. Figures on a GTT are absolutely meaningless unless they are applied to an individual. It makes no sense to assume that every patient's blood sugar must be at a certain level before hypoglycemia can be diagnosed.

But abnormalities cause symptoms—at least sixty-five of them in hypoglycemia—and symptoms indicate the presence of a disorder or disease. A disorder diagnosed by means of a clinical test cannot be treated successfully by a therapist. Nevertheless, when a doctor doesn't see a certain number on a glucose tolerance test, and his patients are suffering from such symptoms as nervousness, tiredness, emotional problems or irritability, he sends them to a psychiatrist. If a patient has hypoglycemia, therapy will not help until the physical problem has been treated. Many of my patients were in therapy for several years before we uncovered their hypoglycemia.

Essentially all disease and illness is man-made and results from the errors of living. We all partake of those mistakes in one way or another as did our ancestors. We may have inherited many things from them, including a deficient body chemistry. Errors of living include smoking, drinking, drugs (from doctors or off-the-street), caffeine, not enough sleep, insufficient exercise, emotional stresses and tensions, traumatic injuries such as accidents, surgery and poor diet.

We cannot escape heredity or many of the stresses of day-to-day living. We may not be able to avoid accidents or some surgery. But we can choose what goes into our stomachs. By eating right, much needless suffering can be avoided.

Part 3
Managing Hypoglycemia

9/Eat Right and Save Your Life

by Wilbur D. Currier, M.D.

As a mother, Jane felt like a failure. Constantly tired, prone to overdisciplining her daughter, and suffering wide mood swings from depression to elation, she finally attempted to take her own life. In desperation, she sought psychiatric help, but still found no relief.

An operation may have saved her from being committed to a mental hospital. Following goiter surgery, she began to have convulsions at night. After the first one, she went to an endocrinologist who thought she might have a brain tumor. He referred her to a neurologist who administered several tests. After finding no evidence of a tumor, he took a closer look at Jane's symptoms and concluded that she could be suffering from hypoglycemia.

It wasn't until after she had two more convulsions in the middle of the night that the endocrinologist became convinced that Jane might have a sugar problem.* He then gave her a glucose tolerance test. It was

---

*The blood sugar is usually at its lowest when the stomach has been without food for several hours; hence the nighttime convulsions.

positive, and he placed her on a hypoglycemia diet. Jane followed it faithfully, and within three months experienced marked improvement in her physical and emotional health.

Hypoglycemia is a form of malnutrition, and its victims have a physical difficulty that must be dealt with and treated daily, if they are to live normal lives. A good nutritional program is essential for any individual to live healthily, but it is vital in correcting an upset metabolism caused by faulty body chemistry.

Jinny has explained that the chief food to the brain is blood sugar. If, for some reason, the food we eat is *not* being turned into that substance and our brain is being starved, serious consequences can result. In the base of our brain we have compact masses of cells called brain centers. These brain centers control our heartbeat, appetite, thirst, hormone production, temperature, sleep, perspiration and, to a large degree, digestion and absorption. They also control all of our senses: hearing, seeing, touching, smelling and tasting.

Is it any wonder that when our brain is being starved we begin to develop all kinds of problems?

If we supply our brain, and the rest of the body, with nourishing food, we can usually correct the body chemistry and get the metabolism back into proper working order, thus avoiding much disease. With food, a good principle to follow is: "If man made it, don't eat or drink it." Conversely, if our food comes directly from Mother Nature, it is almost always good for us, and we can have nearly as much as we want.

Actually, we should try to emulate the caveman in our diet. His foods were meat (including fish and fowl), vegetables, fruits, whole grains and cereals, seeds and nuts. Sugar, as we know it, didn't exist. Contrary to what some people believe, we do not need sugar in our food. People lived without it for centuries and had few of the degenerative diseases we have today. The increased incidence of many of today's ills can be traced to the rise of sugar in our diets—heart disease, diabetes, hypoglycemia, cancer, mental illness, even alcoholism, crime and juvenile delinquency.

One of the first mentions of sugar, from ancient India, states that the cane was "bruised when ripe in mortars, and the juice set in a small vessel until concentrated in form like snow." For centuries it was used in this relatively natural state. The problems began in the beginning of the nineteenth century with American technology. With the invention of the vacuum pan, the steam engine, and charcoal, came the building of great sugar refineries. These turned sugar cane into pure white crystals which had absolutely no nutritional value.

Thus, by the middle 1900s a new disease had entered the world—sugar diabetes. Some scientists contend that diabetes existed for thousands of years, but there is little evidence to support this theory. Dr. G. D. Campbell, a South African expert on the disease, wrote:

> I find it hard to explain why Hippocrates never described a case of diabetes. Such a careful clinical observer could hardly have failed to recognize its...manifestations....Certainly it must have been an uncommon disorder.[1]

Despite the fact that no one knows when sugar diabetes was first recognized, medical science is aware that its incidence increased during the nineteenth century—after sugar refineries came into being. Then came the discovery of insulin which made diabetes more manageable for its sufferers, but did little to stop its devastating effects or the amount of deaths resulting from it.

Dr. Frederick Banting, the discoverer of insulin, tried to tell the people of America that his discovery was not a cure, but merely a relief from symptoms. He suggested that the way to prevent diabetes was to cut down on the use of sugar. In 1929 he wrote:

> In the United States the incidence of diabetes has increased proportionately with the per capita consumption of sugar. In the heating and recrystallization of the natural sugar cane, something is altered which leaves the refined product a dangerous foodstuff.[2]

Dr. Seale Harris also warned that the "excessive use of processed carbohydrates" is one of the causes of hypoglycemia and, eventually, diabetes. Unlike Banting, however, Dr. Harris had no miracle product to help the millions of sufferers. Instead, he cautioned victims to eliminate sugar-saturated coffee, sugar-laden deserts, candy, soft drinks, white flour, white rice and other worthless and harmful food" from their diets. But most physicians ignored Dr. Harris or attacked him for his theories.

Since not much was known about good diet in those days, Dr. Harris and Dr. Banting were ahead of their time. Today nutrition is a science. A few medical schools include it in their curriculum, and more and more doctors are considering poor nutrition as a cause of disease. A few years ago the Heart Association acknowledged that too much sugar may be the cause of some heart trouble. And in 1982, the Cancer Society announced that improper diet may lead to cancer.

My patients with abnormal sugar metabolism must change their eating habits. I don't like to use the word "diet" because today it has a negative connotation—

something unpleasant one must go on to lose weight. Most popular diets are fads, which means they are of a temporary nature and frequently injurious to health. A good nutritional program, however, is beneficial to anyone.

Foremost on any nutritional plan is the exclusion of processed carbohydrates — refined sugars and starches. We need carbohydrates, but should get them from such foods as fruits and vegetables. Our systems do not know how to handle those which have been processed by man, and trouble develops when we ingest them. It is impossible in this country to obtain sugar which has not been processed, be it white, brown, powdered, turbinado, or raw.

Most of us would not think of going to the sugar bowl and eating forty, fifty, or sixty teaspoons of sugar a day, but the average American eats that much disguised in his food or drink, and thinks nothing of it. Not only does this excessive amount of processed carbohydrates cause various diseases including hypoglycemia and diabetes, it also results in hidden allergies. An allergy to sugar is common, but often overlooked.

The next most important factor in nutrition is fats in the blood. Americans get about thirty-five percent of their calories in fats. Some receive as much as forty, fifty, or sixty percent. This is dangerous. Fats are essential to our lives; we would die without them. But try as hard as we can to eliminate them, we still get too many. The fats we do consume should come from whole, natural, highly nutritious foods.

I suggest several ways for my patients to cut the fat from their diets. First, since most of the harmful fats come from animal sources, I advise them to get much of their animal protein from seafoods. Man can't get at fish to inject them with hormones or to feed them improperly because they are part of the natural chain. Second, eat fat-free fowl. Remove the skin of the chicken and cut away the large pockets of fat underneath. This may take a little extra time, but throwing that fat into the garbage is much better than putting it into our families' stomachs. And a final suggestion for cutting down on fat is to use skim, or at least low-fat, milk.

If I limit my patients to the amount of eggs they can eat, it is one a day. Eggs represent one of the most nutritious foods on earth, and I hesitate to restrict the use of them, even though the individual's cholesterol may be a little high. It is a mistake to believe that one or two eggs a day can cause permanent, harmful elevated amounts of cholesterol. What the body chemistry does to the cholesterol after we eat it is the important factor. Abnormal amounts in our systems are caused, not so much by what we put into our stomachs, as by faulty metabolism. If our body chemistry is perverted, problems will surface regardless of what we eat.

The ideal nutritional plan is essentially an ovo-lacto (egg-milk) vegetarian one. A little animal protein is acceptable, but not more than three or four times a week.

Even the following low-fat, no refined carbohydrate nutritional program is admittedly a compromise to an ideal diet. Most of us choose to live in cities where someone else grows our food, transports it, and unfortunately, processes it. When we eat with friends and relatives in restaurants, we actually have no idea how the food has been grown or what has been added to it. Thus, it is difficult to achieve an optimal nutritional intake. We can, nevertheless, strive to eat wholesome foods, such as those suggested in the following food plan:

Low-Fat, No Sugar,
No Refined Carbohydrate Food Plan

Foods Allowed

Very lean meat and fowl (fat free, not more than two or three times per week), seafoods (no oil-packed).

Dairy products, eggs (only 1 daily), certified raw non-fat milk, if available. (May use whole milk if underweight), butter (*very* small amounts), skim milk cheese such as cottage cheese, string cheese or hoop cheese, yogurt and custards.

All vegetables and fruits not listed below.

Banana daily or whole citrus fruit (preferable to the juice).

Raw nuts (preferable to dry roasted) and nut butters (unless weight or cholesterol is a problem. Excellent between meals).

7-grain stone ground bread, 7-grain cereal to cook, or 7-grain granola-like cereals.

Herb teas (if desired).

Whole soybean products.

Artichoke or spinach macaroni and spaghetti (can be obtained in health food store and some supermarkets).

Bread made of oat, soya, high gluten, sprouted grain bread, but 7-grain stone ground bread preferable. No "enriched" flour.

Foods to Avoid

All sugars, refined flour (junk food).

White rice, packaged cereals. (Whole grain rice, whole grain natural cereals permitted.)

Canned fruit, especially sugar packed.

Pie, cake, pastries and candies.

Coffee and tea (may use herb tea).

All alcoholic beverages.

All animal fats and especially cooked animal fats; no sausage, bacon, weiners, fatty sandwich meats, hamburger (but *very* lean ground round, rarely).

Breakfast

Any whole fruit (preferable to the juice).

One egg or *very lean* meat. If cholesterol is too high, eat less meats and eggs.

Pat of butter (small) occasionally with slice of bread as above. Patients to have *no* butter or margarine if blood fats high.

Skim or low-fat milk, or herb tea. Decaffeinated coffee.

In place of the above breakfast, occasionally substitute the Currier Cocktail breakfast eggnog. It is especially good for between-meal snacks.

Currier pancakes. See *Appendix D* for recipe.

10:00 a.m. Snack (choice of)

Skim milk or piece of fruit.

Gelatin (without sugar) mixed with fresh fruits or vegetables.

Raw nuts (preferable to dry roasted or processed, but few if weight or high blood fats are a problem).

Nut mix (nuts, pumpkin seeds, sunflower seeds, raisins, mixed one-fourth each).

Protein wafers or tablets (health food stores).

Glass of Currier Cocktail. See *Appendix E* for recipe.

Lunch (choice of)

Soup (starch and fat free). Let soups cool in refrigerator, skim off all fat, then reheat.

Very lean meats or skim milk cheese or:

Hot lunch consisting of *very lean meat* or seafoods and vegetables.

Skim milk or other allowable beverage, allowable cheese.

vegetable juices.

Fruit, vegetable and fish salads with low fat cottage cheese.

3:00 p.m. Snack

See 10:00 a.m.

Dinner

Soup (starch and fat free).

Very lean meat but preferably seafoods.

Low starch vegetables, salad (chop up *many* raw vegetables).

Milk (low-fat).

Fresh fruit.

Bedtime (choice of)

Low-fat milk, cheese or 1 glass certified raw, nonfat milk if available).

Fresh fruit.

7-grain stone ground bread. Nut butter for spread (if calories needed and if blood fats are low).

During the night ... when one awakens and cannot go back to sleep. (choice of)

Milk or unsweetened orange juice.

Low-fat cheese or a sandwich made of allowed bread and *very* lean meat.

Nuts of any kind (preferably raw).

Protein wafers, tablets or protein drink. No refined carbohydrates or sugar.

Nutritional Supplementation

A good all around vitamin-mineral tablet morning and evening.

Vitamin C, at least 1000 mg. or more daily.

Lecithin capsules morning and night (no-oil lecithin capsules).

Other supplementation will be recommended as indicated by Computerized Diet Survey and other studies.

Stop smoking

Stop Alcoholic Beverages

We cannot overemphasize the importance of this program. It essentially consists of the strict elimination of rapidly absorbed carbohydrates (especially sugars, white flour foods and white rice) in order to eliminate the sudden rise in blood sugar with its subsequent rapid fall. This is not a strange or unusual diet, but rather one that practically all people should follow. It concentrates on the most nutritious foods and eliminates harmful ones.

A hearty breakfast and between meal feedings are advised to prevent slackening off of blood sugar levels, which are prone to occur three to four hours after eating. The hypoglycemic especially should have six snacks a day instead of three meals. Salt is allowed in unrestricted amounts because of the tendency to sodium depletion, unless a person has certain heart or kidney diseases. Particularly during hot weather, supplementary salt in the form of tablets is advised to replace the loss caused by perspiration and other metabolic processes.

No *specific* treatment can be prescribed for a hypoglycemic patient without the thorough diagnosis and work-up as described in Chapter 8. However, the following suggestions are ones which would be advantageous for anyone to follow.

First, the diet should be as nearly one hundred percent free from processed carbohydrates as possible. Fats should be no more than fifteen percent in calories of the total diet. Foods should be as nearly free of additives of all kinds as possible. Foreign substances, such as chemical additives, in our stomachs pervert metabolism by harming and destroying enzyme processes and damaging intracellular chemistry. They probably adversely affect all tissues of the body, but definitely the delicate, highly specialized tissues of the brain, glandular system, and other hormone-producing tissues. Animals fed even small amounts of food additives become sick and deteriorate seriously.

Second, almost everybody needs vitamins and food supplements. It is virtually impossible in these days of chemical fertilizers and polluted water and air to obtain completely nutritious foods from the soil. I recommend a high-potency all-around general vitamin-mineral preparation in which the minerals are chelated. Ordinarily, when we take a mineral tablet or absorb natural minerals from food, the body attaches an amino acid to aid digestion. Many body systems do not produce the proper amino acids, however. Chelated minerals take care of this problem because they have the acids already attached so that the minerals can be easily assimilated, even though the patient's body process is not functioning properly.

I also recommend taking from two thousand to ten thousand milligrams of Vitamin C each day. Tablets are the most widely accepted form, but I advise powdered Vitamin C in case of emergencies, such as having been exposed to cold germs or having a particularly stressful period of constipation. The powder is absorbed much faster and is effective much sooner than the tablets. It can be taken in any kind of unsweetened fruit juice.

Vitamin E is another vitamin recommended for all my patients. Volumes have been written about Vitamin E, but suffice it to say here that it utilizes oxygen more efficiently and protects against smog and toxic chemicals.

The hair test may indicate that an individual needs

extra minerals, and the gastric analysis and other tests may reveal the need for digestive enzymes. The use of other vitamins and food supplements depends upon the findings of the complete work-up.

Third, this program is entirely different from the diabetic diet. A diabetic must count calories, grams of carbohydrates, and carefully plan exchanges, one for the other. The hypoglycemic diet is much simpler. One can eat as many of the allowable foods as he or she desires, must stay away completely from the unallowable foods, and eat frequently. Eating often may be a stumbling block to many people, especially those who are overweight. However, it must be understood that these continual feedings are necessary for correcting the metabolic dysfunction. Fluctuation of blood sugar levels must be avoided. One way to do this is by keeping the system supplied with nutrients. This tends to even the sugar levels off (see diagram, page 118) and bring the whole metabolic system back to normal. When this is achieved, fats will be utilized as needed, stored in the proper place, or eliminated altogether. Weight will return to normal. Because of this, those who are underweight will also utilize fat properly and begin to gain weight.

Although it may sound so at first, this nutritional program is not boring or uninteresting. Fresh vegetables, steamed until they are just tender, with a little butter and salt, are delectable. Fresh fruits, frozen with no sugar, and custards made with a little sweetener can take the place of pies and cakes and other sugary deserts. Jellies and jams can be canned with pineapple juice instead of sugar. And whole grains and cereals are delicious when soaked overnight in apple, pineapple or other unsweetened fruit juice.

It is surprising how painless it is to get sugar completely out of your life. Amazingly enough, once you start eating good wholesome foods, your "sweet tooth" will disappear, and eating will become a pleasant but challenging experience.

10/The Hassles in Getting Good Food

"**Y**our whole grain bread is so good!" I said cheerfully to the girl behind the cash register. "That's the kind all restaurants should serve, instead of the white junk they usually have."

"That's our specialty," she beamed. "We make it right here. And I think the honey butter we put on it really helps, too."

I had gone to that restaurant because the home-baked breads were not only delicious, but nutritious. Having been on the hypoglycemia diet for several months, I was not eating anything made with white flour or sugar. But here, unknowingly, I had eaten something just as bad for me, maybe worse—honey. Since one teaspoon of sugar could put me in bed for a day or longer, I hoped the honey wouldn't do the same.

The more I thought about that cashier's remark, the more frustrated I became. What about all the diabetics and the other hypoglycemics who eat there? Is there no way they can be sure that the foods they eat will not further upset their metabolism? We Americans have a built-in addiction to sweets, and catering to that habit is good business, whether by a food manufacturer or a chef.

We can buy little food these days without some kind of additive, especially if it comes in a box, a can or a package. These chemical additives are not good for anyone, especially a person who already has metabolic problems. A hypoglycemic must be especially careful about additives which can worsen his condition.

The Food and Drug Administration (FDA), the federal organization set up to protect the nation's food, defines an additive as:

> Any substance the intended use of which results or may reasonably be expected to result, directly or indirectly, in its becoming a component or otherwise affecting the character of a food.[1]

Today, some 2,800 substances are intentionally added to foods to produce a desired effect. As many as 10,000 other compounds or combinations of compounds find their way into various foods during processing, packaging or storage.[2]

Are these chemicals necessary? If we're talking about the importance they play in profits for food manufacturers, the answer is yes. If we're referring to their role in consumer health, the response is a resounding no.

Dr. Harold Stone, president of the International College of Applied Nutrition, writes:

> Unfortunately, no man can duplicate in the laboratory the necessities of the body as they are supplied in proper food. We can perhaps avoid some deficiencies by taking concentrated vitamins or minerals; this is only an aid—nothing more.[3]

Michael Jacobson, Ph.D., director of the Center for Science in the Public Interest, writes in his book,

Eater's Digest:

> One of the oldest ways that merchants have of enhancing profits is to substitute an inexpensive ingredient for an expensive one.... It is unlikely that behind the closed boardroom doors of this nation's food and chemical companies, the managers are grappling with the problem of how to solve once and for all America's problems of hunger and malnutrition. More likely they are calculating how they can increase sales and asking how food additives can help.... Food additives help companies make money and that, in a nutshell, is why additives are usually used.... Soft drink bottlers promote artificially sweetened drinks because saccharine is far cheaper than sugar on a per bottle basis, not because such drinks help many persons lose weight; profit, not effectiveness, is the impetus.[4]

The trade magazines also substantiate this philosophy. One ad puts it this way:

> If the high cost of traditional ingredients is milking your profits, switch to these quality non-dairy products. They're far more economical to buy. They need no refrigeration and they have longer shelf life. So start saving with our *(product name)* today. Our food ingredient does the job for about half the price of the traditional products.[5]

Another company advertises in the same magazine:

> You can make a fortune in convenience foods.... Our specially formulated binders turn out tender, juicy meats every time.[6]

Still another proclaims:

> Replace (fresh dairy cream) with (our non-dairy cream)...(it) successfully replaces a whole series of other ingredients in your formula. This saves you time, trouble...and money.[7]

But this is the best additive ad of them all:

> Ham it up for 16¢ an egg with new *(product name)*. They look like ham. Eat like ham. Taste like ham. But at 1¢ a serving, they sure don't cost like ham. Now you can add ham-like flavor and texture to scrambled eggs, omelets, pancakes, waffles, biscuits, casseroles, western sandwiches — for a penny.[8]

By 1906 the public outcry against the "food tainters" was so great that President Theodore Roosevelt signed the Pure Food and Drug Act and established the Bureau of Chemistry of the Department of Agriculture. The first director of the bureau was a physician who also was a chemist. In the early days, when a new chemical was developed as a food additive, it was tested by the bureau. If, after testing, it was ruled that the chemical would adulterate the food for which it was intended, the additive was prohibited. The agency was indeed a "watchdog" over America's food.

That first board soon became the Food and Drug Administration and the commissioner became a political appointee who sometimes knew nothing about foods and drugs. The 1958 Food Additive Amendment took the burden of proof from the watchdog agency and gave it to, of all people, those it was watching — the food manufacturers!

Additives are no longer tested by the government agency set up to make sure that we do not get impure or inferior foods. Today the FDA has nothing to say about the *quality* of food being manufactured or the total number of additives that can be used. FDA regulations state only that additives used in the processing of food must be proven safe for consumption by human beings.

The FDA does not set nutritional requirements and has no authority to require manufacturers to produce food that is nutritious. As long as an additive does not harm laboratory animals within a specified time, as determined by the manufacturer or by a food manufacturer, it is approved by the FDA.

The Food Additive Amendment was responsible for a classification of additives which is still in use today — the Generally Recognized as Safe (GRAS) list. When that amendment was passed, hundreds of chemical additives existed which had been in use for several years. Because they had not been known to harm anyone, they were informally, not scientifically, classified GRAS. Most chemicals which were and still are on the GRAS list never went through a pre-marketing testing procedure. About eight hundred substances were on the original list. This means that unknown quantities of untested chemicals are in our food — even baby food — today.

One GRAS substance being used extensively in baby food is modified tapioca starch. Tapioca is a nutritious food but when an element is modified, it is altered chemically. Because modified starch is supposedly easier to digest than natural starch, it is put in baby food. But infants do not have the resistance to chemicals that adults do, and these substances may be doing irreparable harm to their little systems. No one knows how much residue is going into our babies' stomachs.

I gave my children baby food twenty-five years ago because I didn't know any better. I felt proud when they eventually grew out of it and had their first finger foods. But little did I know that those hot dogs might have contained ground-up bones or residues left from steaks and chops. Other ingredients include water, salt, sugar, flavoring, and sodium ascorbate (vitamin C). Although that last ingredient might seem to indicate that the hot dog is nutritious, when vitamins are added to food, more natural ones were destroyed during processing than were put in synthetically.

Canned soup became a staple in my children's diet after they graduated from baby foods. I guess I didn't believe all the ads that said they were nutritious, because I made them all with milk instead of water, as the directions called for. Milk at least added some nutrition. What I didn't realize was that many soups on the grocers' shelves contained and still contain monosodium glutamate. Now, there's nothing wrong with MSG, except that it has been shown in some laboratories to "cause brain damage in young rodents and brain damage effects in rats, rabbits, chicks, and monkeys. (It is) on the FDA list of food additives needing further study for mutagenic (producing change in chromosomes or genes), teratogenic (causing serious malformations), subacute (having severe symptoms), and reproductive effects.... Females treated with MSG had fewer pregnancies and smaller litters, while males showed reduced fertility."[9]

There is no way of telling how many people MSG has affected adversely in the last few years. It was held responsible for the "Chinese restaurant syndrome" several years ago, when individuals who had eaten in Chinese restaurants suffered chest pains, headaches and numbness. No cause was ever found, except that most Chinese restaurants use MSG profusely in their foods.

Baby food manufacturers stopped using MSG because of public pressure. Too many people considered it injurious to infants. I wonder if it is any less harmful to older children and even adults who use it almost daily?

Another common item in my children's diet was dry breakfast cereals, until I noticed that I would be hungry myself an hour after eating a bowlful. Consequently, I never let them start the day with just a bowl of cereal. Letting children get their own breakfasts by fixing a bowl of cereal is, I believe, one of the worst practices in America.

In fact, several years ago I read that there is as much nutrition in most boxes of cereal as there is in the cereal itself. That's not quite true; *some* cereals do have a little food value.

There are good cereals on the market, but many are heavily sugared and contain preservatives and other chemicals. Some have been classified as candy and are not the proper food with which to start the day. Breakfast is the most important meal and should consist of such nutritious foods as eggs, whole wheat or whole grain bread, whole grain cereal, and fruit or juice. The mostly-sugar-and-chemical fruit "drinks" should be avoided. Some doctors and nutritionists believe that no category of food on the market is more expensive, more profitable, and more nutritionally disastrous than most breakfast cereals.

If you desire to prepare a natural, wholesome, nutritious and good-tasting cereal for breakfast, try the following:

> At a health food store, choose four or five whole grains from their variety, which will usually include whole wheat, rye, rolled oats, brown rice, bran, chia, wheat germ, sunflower seeds, and sesame seeds. You

will also have a broad selection of nuts from which to choose one or two – almonds, filberts, pecans, walnuts.

Chop the nuts fine and mix with the grains and seeds. Store in a cool, dark place, preferably in the refrigerator. The night before you want to serve this delicious cereal, take at least one half cup per person, cover it with the same amount of water, and let it stand overnight in the refrigerator. The next morning it will be just the right consistency for eating.

For hot cereal, simply warm it on the stove or heat it in the microwave a few seconds.

If overnight preparation is not feasible, the same quantities can be cooked for four or five minutes. You'll have the same delicious meal, but it won't be quite as nutritious because any food loses nutriments when heated to high temperatures.

This cereal can be sweetened with a little honey unless a diabetic or hypoglycemia diet is being followed. If all forms of sugar are being avoided, soak or cook the cereal in unsweetened fruit juice.

You won't be hungry a half hour after eating this homemade cereal for breakfast; it will satisfy until lunch! It costs only a few cents and is one hundred percent nutritious. Some processed cereals, on the other hand, are almost ninety percent refined sugar and starch and contain more artificial flavorings than nutriments. They are mixed with hydrogenated fats and oils and preserved with chemicals that are harmful to our systems.

Potato chips are another unnutritional food, though not as devastating as cold cereal because they are usually eaten along with other foods. Heated to an extremely high temperature in perhaps two or three kinds of oil, they don't at all resemble the original nutritious potato. Although the addition of chemicals and preservatives makes them more profitable to the manufacturer by providing a longer shelf life, the additives do nothing for the benefit of those who snack on them. One popular brand of chips has "no preservatives" printed in large red letters on the front of the package. But one must read the fine print to learn that the product contains *MSG,* artificial flavors, sugar and several other chemicals.

Ice cream also is an attractive food with little nutritional value. Commercial ice cream usually is nothing but sugar, artificial flavorings and colors and other chemicals. Even some of the "natural" ice creams contain chemicals and preservatives. About the only way to have good ice cream is to make it "from scratch" with wholesome ingredients.

The danger of additives is not only in the amounts used, but also in the combined effect of *all* the chemicals we ingest. Statistics show that the average American eats more than ten pounds of chemical additives a year. New ones are being manufactured all the time, and there are no regulations concerning how many can be put into our food. No one really knows what effect one additive may have upon another.

Ross Hume Hall, professor in the department of chemistry at McMaster University in Hamilton, Ontario, Canada, observes:

> Toxicology has made a slight step toward recognizing multiple agents. The National Academy of Science Committee notes that two or more additives can act synergestically to promote high tumor-yield in animals, whereas separately they do not. Few of the several thousand compounds on the GRAS list have been tested in animals for carcinogenicity (causing cancer). If the FDA were to start testing the possible combinations of even just two or more chemicals,

Reprinted by permission of Tribune Company Syndicate Inc.

all the biologic resources in the country could not cope adequately with carrying out these tests. This fact alone demonstrates the poverty of their approach to evaluating food safety. Classic toxicology has to be scrapped, and a completely new approach must be devised. A new approach must take into account the multiplicity of interactions in the total food service environment, from the mode of agriculture to food fabrication and its distribution to the consumer.[10]

William C. Hale, senior project manager of the food and agribusiness section of Arthur D. Little Inc., said in 1976 that no one is quite sure what the public is eating anymore. About the same time, the president of Morton Foods was criticized because his company was making lemon cream pies which contained no lemon, no cream, and no eggs. His response was amazing. He argued that the Morton Company was doing society a favor by putting artificial ingredients into their pies because this was getting the public ready for the future when all foods will be artificial.

In the early 1970s I knew enough about nutrition to realize that food additives were not good for me or my family. I held to the principle, "don't eat it if you can't pronounce it." In the last decade, so many more chemicals are being used in food that now I stay as far away from pre-packaged or prepared foods as possible.

That's fairly easy for me to do, since my children are grown, and I am responsible only for what goes into *my* stomach. But avoiding prepared foods is practically impossible for many families, especially those with working wives, mothers and single parents. Food additives are here to stay, and we'll never go back to the "good ole days" when most food was eaten in its natural state, home-canned, or otherwise preserved without chemicals. Yet something needs to be done to guard the nutritional aspect of our foods. Things can be changed, but it might take an act of Congress.

In the area of additives, consumers, both individuals and groups, can make themselves heard by writing their congressmen. Laws are needed to keep new products off the market until their safety is assured by a neutral agency and to withhold substances from the market once their potential harm has been declared.

The "interim" status of additives should be removed. Under present law, harmful substances can be in food for much too long a time. For example, it took the FDA sixteen years, from 1954 to 1970, to ban cyclamates, even though much evidence that they caused cancer was available. Saccharin has an even gloomier history. Considered a dangerous chemical in 1907, it was banned by the first president of the Bureau of Chemistry, the forerunner of the FDA. However, a commission was appointed to re-evaluate the sweetener's safety. It was decided that the ban wasn't necessary, and saccharin went back on the market. One of the members of that commission was Ira Remson, the man who discovered saccharin.

In 1977 Canadian scientists announced their findings that saccharin was causing bladder cancer in rats. The product was banned in Canada, and the FDA decided to ban it here. But protests from thousands of saccharin users, mostly diabetics, and loud cries from the chemical companies and food manufacturers who were profiting greatly from the use of five million pounds a year, prevailed. Instead of upsetting the FDA's regulatory system by permitting the use of saccharin, Congress passed a law that exempts it from food additive regulations.[11]

Since then, further Canadian studies have proved conclusively that saccharin causes cancer in human beings as well as in animals. Other studies have shown that persons who use the sweetener are 170 percent more likely to develop cancer than those who don't use it.[12] But it is still being used freely in the United States—74 percent of it in diet soda, 14 percent in dietetic food, and 12 percent as a tabletop replacement.[13]

Food colorings is another item which has given the

FDA problems. When the first legislation for food dyes was passed in 1906, only seven out of an existing eighty, when tested, were shown to be composed of ingredients which demonstrated no harmful effects.

In 1938 the regulations were changed; fifteen colors were certified at that time. Twelve years later, three were de-listed because the candy and popcorn in which they were used made children ill. Since then, nine others have been removed, but others have been added.

Food colors are used solely to change appearance and contribute nothing to nutrition, safety or processing. Added to almost all processed foods, their use is not restricted, except that they cannot be used to cover up unwholesome products. Artificial colors must be listed as ingredients in all foods except butter and cheese.

Should history repeat itself, any food dye now in use could prove harmful and be taken off the market in a few years. Even now, one is highly suspect—Yellow dye No. 5, a coal tar derivative. According to the FDA, it causes "allergic reactions—mainly rashes and sniffles—in an estimated fifty to ninety thousand Americans."[14] Although efforts were made to ban this color, aspirin-sensitive persons have been reported to develop life-threatening asthmatic symptoms when ingesting it.[15] However, it is still being used in about sixty percent of both over-the-counter and prescription drugs, as well as in many foods.

In 1966 the FDA proposed to limit the use of Yellow No. 5 in food. The color industry objected strongly and petitioned for a permanent unlimited use, and the FDA granted their request.

In 1977 a consumer advocate group petitioned the FDA to prohibit the use of Yellow No. 5. That appeal, made on behalf of the consumer, was denied, but the

FDA did rule that the coloring must be listed in the ingredients of any food product that contained it. Now, all persons with allergies need do to make sure that they are not getting Yellow No. 5 in their food is to read labels on all packaged foods they buy.

Dr. Franklin Bicknell, writing in the *British Medical Journal*, says that food shouldn't have to be dyed. He argues that it is virtually impossible to be sure of the safety of any synthetic colors in use today. Some of them, he contends, may have cancer-causing properties and therefore, all should be banned.

Why are dangerous substances allowed in our food for even a day, let alone years? Something is terribly wrong when a chemical known to be potentially life-threatening is allowed to be used. Dr. Currier says that toxic materials injure the immune and endocrine systems and will accentuate disease. "The metabolism," he asserts, "is fragilly balanced, particularly in persons who have genetic tendencies toward metabolic imbalances. Hypoglycemics must be particularly careful about ingesting foreign substances."

In 1969 the President's message to the first session of the 91st Congress of the United States related entirely to the protection of consumers' interests. In referring to the nation's food supply, he spoke of the "Buyer's Bill of Rights" and asserted that:

> The buyer in America today has the right to make an intelligent choice among products and services. The buyer has the right to accurate information on which to make his free choice. The buyer has the right to expect that his health and safety is taken into account by those who seek his patronage. The buyer has the right to register his dissatisfaction, and have his complaint heard and weighed, when his interests are badly served.[16]

It would seem that our interests are being badly served by the food manufacturers in this country. But as things stand now, we have no choice as to what chemicals are going into our food.

In that same message to Congress, the President ordered a full review of food additives and an examination of the safety of the GRAS substances. The FDA contracted with the Federation of American Societies for Experimental Biology (FASEB) to provide expert opinions on the safety of substances. FASEB's Select Committee on GRAS substances (SCOGS) did the review on 469 selected items. Some evaluations took only one or two years, others as long as six. Each substance studied was placed into one of the following five categories:

1. Considered safe. Remains on GRAS list.
2. Continues on GRAS list with limitations on amounts used.
3. "Interim" status; uncertainties exist requiring further study.
4. Evidence insufficient to determine that adverse effects reported are not harmful to public health. Recommend establishing safer usage conditions or remove ingredient from food. (No time limit).
5. Insufficient data upon which to base an evaluation.

Of the substances studied, 407 were placed in classification one or two and are still generally recognized as safe. Uncertainties existed with twenty-three which were put in category three or "interim." They can be used freely until those doubts are cleared. Five were placed in the fourth classification—insufficient evidence to say that they are not harmful. It was recommended that safer usage conditions be established for all five, or that they be banned. The recommendation was made in 1979, but no action has been taken. The remaining thirty-four are in the fifth classification—insufficient data to make an evaluation.

In other words, in the thirteen years since the President's order, no conclusion has been reached concerning thirteen percent of the items studied, and they are still being added to food. Furthermore, about four hundred GRAS items have had no review whatever.

It is interesting to note also that this review of additives relied mostly on studies already made by food manufacturers, not on actual tests.

The 1977 *Dietary Goals for the United States*, prepared by the Select Committee on Nutrition and Human Needs of the United States Senate, gave the following warning:

> Although food additives as a category may not justifiably be considered harmful, the varying degrees of testing and quality of testing and the continuing discoveries of apparent connections between certain additives and cancer, and possibly hyperactivity, give justifiable cause to seek to reduce additive consumption to the greatest degree possible.... In Nutri-Score, Fremes and Sabry suggest that necessity should be the touchstone for the use of additives. They argue, as do others, that only those additives that serve a necessary function should be permitted in food.... It is apparent that necessity most strictly defined has to do with protecting food safety.

In 1979 FDA *Consumer*, the official magazine of the FDA, reported that scientists are constantly searching for safer alternatives to food additives. The same article stated that the scientists will never be able to guarantee the absolute safety of additives. "Ultimately," it explained, "it is up to the consumer to decide what degree of risk is an acceptable price to pay for foods that keep well and are appealing, convenient, and readily available all year round."[17]

As senators urge the reduction of additives, the FDA seems to encourage their use or, at least, does nothing to restrict them. Apparently, the government is no longer the watchdog or protective agency over our food, as it used to be.

Who, then, is responsible for our nutritional needs? No one—except ourselves! It's strictly up to us, and should be a matter of concern to everyone, particularly to those who have hypoglycemia.

11/The Rest of the Deadly Don'ts

Sugar and other processed carbohydrates, some drugs, and food additives are not the only substances which a hypoglycemic must avoid, however. Caffeine, nicotine and alcohol are also deadly don'ts.

In speaking of caffeine, one authority made this statement: "No drug with such potential danger has been so widely used with such little supervision."[1] His is just one of the many voices warning of the great harm in drinking coffee, tea and soft drinks containing caffeine. These products are harmful to everyone who uses them, not just hypoglycemics.

Often we hear the complaints of consumers:

"Coffee keeps me awake at night."

"Coffee makes me nervous."

"Coffee makes my heart beat too fast."

Such comments corroborate the claims of scientists who assert that caffeine indeed causes nervousness, insomnia, irregular heartbeat, ringings in the ears and, in high doses, convulsions. Recent studies show that this drug crosses the placental barrier and causes malformation of the fetus. Because of these reports, women are being counseled by prominent doctors, the March of Dimes, and other concerned organizations to limit their use of caffeine during pregnancy or run the risk of having deformed babies.

Caffeine is even more dangerous for hypoglycemia victims because it alters blood sugar release and uptake by the liver, consequently aggravating the low blood sugar condition. This is why caffeine is one of the substances prohibited in a hypoglycemic's diet. As little as one cup of coffee or strong tea can cause serious low blood sugar problems in a hypoglycemic. Dr. Paavo Airola believes that the combination of coffee and sugar is particularly harmful. Sugar enters the bloodstream quickly and directly, while the caffeine in coffee adds to the total sugar level by acting through the adrenals, brain and liver.[2]

Most adults are aware of the potential harm in drinking too much coffee, but how many of us know that children are consuming just as much, if not more, caffeine in soft drinks? Not only colas, but also other light beverages contain this potent stimulant. According to a survey published in the August 1981 edition of *Nutrition Action,* the average consumption of soft drinks in the United States is forty gallons per person per year. This fact has removed soda pop from the category of occasional treats and made it a great concern of nutritionists, doctors, and others concerned with public health.

Do parents who buy decaffeinated coffee for themselves, yet load their shopping carts with cokes and other beverages, realize how much caffeine they are giving to their children? A twelve-ounce can of coke or other caffeine-containing soft drink has about a third as much caffeine as a cup of coffee. But because a child weighs much less than an adult, the caffeine in that can of pop is equivalent to a cup of coffee drunk by an adult.

Independent scientists have advised the Food and Drug Administration to give priority study to the effects of caffeine on children. They urged the evaluation of the impact of continuous consumption of the drug during the period of brain growth and development, since the estimated levels of caffeine intake at those ages are known to cause central nervous system defects in adults.

But Americans are downing large amounts of caffeine in other ways without knowing it. Some popular over-the-counter drugs contain varying amounts of it. One tablet of Dexatrim, for example, contains up to five times the amount of caffeine as a five-ounce cup of instant coffee. Excedrin has the same amount as a cup of percolated coffee, while No Doz has approximately double the amount.[3]

However it is ingested, caffeine travels rapidly through the bloodstream, reaching practically every part of the body within five minutes. It constricts some blood vessels and dilates others. The metabolic rate is increased by ten percent, and in turn, intensifies the output of essential stomach acid and urine. A stimulant to the nervous and respiratory systems, caffeine can cause the muscles of the heart to constrict vigorously.[4]

Physical dependency is another risk that caffeine users take. Dr. Kurt W. Donsbach, well-known nutritionist and author, calls coffee the "national addiction." Persons who consume large amounts of it can suffer withdrawal if they stop suddenly. The symptoms are similar to those experienced by individuals going off any other narcotic – headaches, irritability, lassitude and nausea.

One can break free from the caffeine habit, however, with little or no withdrawals. A good way to cut down quickly on caffeine is to use instant coffee, each time decreasing the amount. Instant coffee has three times less caffeine than percolated coffee. Eventually, the switch can be made to a decaffeinated brand.

Nicotine, another of the deadly don'ts for hypoglycemics, is also injurious to everyone. In the early 1970s the Federal Trade Commission asked Congress for permission to require a warning on all packages of cigarettes which would name specific diseases caused by smoking. This is what they wanted it to say:

> Cigarette smoking is dangerous to your health and may cause death from cancer, coronary heart disease, chronic bronchitis, pulmonary emphysema and other diseases.

A loud cry resounded from the tobacco industry which, naturally, didn't want such explicit and damaging information on their product.

Instead of granting the FTC's request for precise labeling, Congress mandated that the existing warning should be changed from "The surgeon-general has determined that cigarette smoking *may be* dangerous to your health," to "Cigarette smoking *is* damaging to your health."

Because more evidence has surfaced in recent years on the dangers of smoking, many groups and individuals are urging Congress to require specific labeling. Inconsistencies still exist, however, when one branch of the federal government warns against the dangers of tobacco, while another, the Department of Agriculture, encourages smoking by lending tobacco growers as much as 200 million dollars a year.[5]

Smoking is particularly dangerous for those who are trying to control hypoglycemia. Indirectly, the nicotine causes a fall in blood sugar by stimulating the adrenal glands which, in turn, influence the liberation of glycogen from the liver.

Numerous studies show that nicotine affects the brain of a well person in much the same way that it reacts on the pancreas of a hypoglycemic. In the brain, nicotine causes the release of hormones which make the heart beat faster and the blood pressure rise. Just as a bit of sugar, caffeine or alcohol can make a hypoglycemic feel better for a short while before the blood sugar drops, nicotine also makes one feel calmer. But soon, the level of nicotine in the blood drops and the smoker becomes jittery, craving another cigarette to calm down. The cycle starts and a chain smoker is born! Smoking causes craving, and craving causes smoking.

Nicotine isn't the only harmful ingredient in cigarettes. In 1979 Assistant Secretary for Health and United States Surgeon General Dr. Julius B. Richmond became concerned about the chemicals and some of the "natural" substances being added to tobacco. Many of these are safe for human consumption, but there is no guarantee that they are safe *when burned.* The principal flavoring in cigarettes is sugar, which accounts for about four percent of the tobacco by weight in most United States brands. When sugar is burned with tobacco, the tar yield is increased. And some forms of sugar, when heated, produce catechol, the major known cancer-causing agent in tobacco smoke, according to American scientists.[6]

Obviously smoking is not only harmful, it is dangerous, contributing possibly to 350,000 deaths a year in the United States. How, then, can government continue to give price supports to the tobacco industry when so many users are being harmed? Instead, Congress should do everything in its power to limit tobacco sales, rather than encourage them.

Alcohol, as well as caffeine and nicotine, is another item for the hypoglycemic to avoid. In his book, *Diet Revolution*, Dr. Robert C. Atkins wrote:

> When an alcoholic succeeds in getting off alcohol he usually substitutes sweets ... because almost all alcoholics are hypoglycemic and sugar provides the same temporary lift that alcohol once did.[7]

Medical science is much aware that alcohol causes hypoglycemia. But seldom is any connection made between a malfunctioning metabolism, nutrition and alcoholism. The only people who seem to make this association are the nutritionists and doctors who have become nutrition-minded. Drs. E. Cheraskin and W. M. Ringsdorf Jr., with Arline Brecher, authors of *Psycho-Dietetics*, declare that:

> A great many researchers who have lifted the alcoholic off the couch and placed him under the microscope are convinced that uncontrollable drinking is a metabolic disorder that can be treated by nutritional therapy....
> Doctors are, of course, aware of the fact that alcoholics frequently suffer from malnutrition. But ... they assume the malnutrition to be the result of alcoholism, not a contributing factor. Malnutrition of the brain cells (not enough food in the form of blood sugar getting to them, or hypoglycemia) is simply not considered as a *cause* of alcoholism.[8]

Such a close relationship exists between blood sugar levels and alcohol that it just cannot be overlooked.

Dr. Airola contends:

> The relationship is reciprocal. Chronic drinking, just like excessive sugar in the diet, contributes to the development of hypoglycemia; and a person with hypoglycemia is a potential candidate for alcoholism. When he finds that alcohol produces the same effect as sugar, he becomes a compulsive drinker. A dangerous cycle ensues; alcohol improves his sense of well-being temporarily, so it becomes necessary for him to drink most of the time in order to feel comfortable and symptom-free. He has become a chronic drinker, an alcoholic.[9]

Who can say which came first? Does hypoglycemia cause excessive drinking in some people, or does alcoholism cause the low blood sugar? Many authorities believe that the metabolic error is inborn, or exists from birth.

Some studies have shown that alcoholics craved candy and other sweets when they were young, and that they were often tired and depressed — all early symptoms of an upset sugar metabolism. Eating sweets or drinking soft drinks satisfied their craving in youth and temporarily cured their depression and fatigue. As they grew older and discovered that alcohol had the same effect, they became heavy drinkers and eventually alcoholics.

The ingestion of a *small* amount of alcohol, especially by a child or young teenager, can cause a temporary low blood sugar condition. Although this can be corrected easily, the continual drinking of alcoholic beverages causes a chronic hypoglycemia which is hard to manage. It is clear that alcoholics, because they put so much alcohol into their systems, are almost certain to end up with chronic low blood sugar. In Dr. Currier's words, "an alcoholic is a sitting duck for hypoglycemia."

Many articles have appeared in the medical journals about alcohol-related hypoglycemia. Drs. Francis Pottinger, J. D. McAllister, J. J. Smith, Harold Lovell, John W. Tintera, Roy C. Gumpel, and Edwin H. Kaufman were early proponents of this theory. In 1965 Drs. Gumpel and Kaufman wrote:

> The failure to recognize a hypoglycemic origin of bizarre nuerologic signs and symptoms and abnormal behavior in patients who have recently consumed alcohol may result in irreversible cerebral (brain) damage and death.[10]

An article in *Nutrition Review* appeared the same year regarding this relationship:

> Although it is common knowledge that habitual use of alcoholic beverages may cause serious illness, the metabolic derangements which are involved remain poorly understood. In all probability, few physicians are aware of the fact that alcohol can, under certain circumstances, cause severe or even fatal hypoglycemia. Actually, a number of published reports of this condition are to be found, yet the subject seems to have escaped general recognition.[11]

And in 1966 *Aerospace Medicine* printed an absorbing article about alcoholism and hypoglycemia which reported:

> There is recent reference in both medical and aviation literature to the role of alcohol in aircraft accidents, and the medical literature is replete with case histories and discussions of alcohol-induced hypoglycemia; however, the problem of alcohol

> hypoglycemia seems to have escaped general recognition.... We have undertaken an investigation to evaluate the role of alcohol-induced hypoglycemia (AIH) in the 30 percent of fatal aircraft accidents in the Southwest Region which are associated with alcohol ingestion.[12]

Although controversy surrounds the connection between alcoholism and hypoglycemia, much evidence points to a close affinity between the two. Whether alcoholism causes hypoglycemia or vice versa, treatment for an upset metabolism should be a part of every alcoholic's rehabilitation. One writer says that hypoglycemia occurs frequently enough in chronic alcoholics that all who go to emergency rooms with seizures or altered states should have their blood glucose routinely determined.

Controlling alcoholism would be much easier if hypoglycemia were brought under control first. But this approach has never been taken by most of the well-intentioned people and groups who are familiar with alcoholism and unaware of the connection between the two.

Clearly, caffeine, nicotine and alcohol are three deadly don'ts that hypoglycemics can ill afford. Their consumption only aggravates a low blood sugar condition already initiated by foods saturated with sugar and additives.

12/Rime, Reason and Rescue

As we have seen, hypoglycemia is an insidious illness that affects millions of Americans. Its devastating symptoms not only torment sufferers, but drastically affect the lives of their families and friends.

Low blood sugar is wrecking the lives of its victims, often without their knowledge. One wonders how many seemingly mental, emotional and spiritual problems can be attributed to this condition. Writing in the March 17, 1983, issue of *Science News*, Joan Arehart-Treichel stated that "since no comprehensive scientific study has been conducted to find out, the answer depends on which physicians you talk to.... Until (such) a ... study is made to get the answers, no physicians, however blatant or ardent, will have the final word on the subject."

Because of the controversy raging over this metabolic disorder, finding the cause of the trouble is, in many cases, so difficult that it can take years. We have sought to unravel the complex nature of the problem and to help shorten the time it usually takes to detect this often illusive ailment.

While examining both sides of the controversy, we have given documented evidence to show that hypoglycemia can and does cause depression, chronic fatigue, allergies, nervous breakdowns, alcoholism, drug abuse, inadequate sexual performance and a host of other difficulties. Although the debate has raged for decades, more and more physicians are beginning to recognize low blood sugar as a metabolic dysfunction and nutritional problem.

Dr. Jerry Earll, an endocrinologist and chief of internal medicine at Georgetown University Medical Center in Washington, D.C., is among them. Late in 1982 he declared that hypoglycemia is now being recognized by endocrinologists as a genuine systemic dysfunction and not the neurotic meanderings of difficult patients. He offered supportive information on why hypoglycemia is so often overlooked:

> There is no way to make a definition that will apply to everyone. One patient may have symptoms with blood sugar at 40 and another may have no symptoms at 20. Hypoglycemia has not been recognized in the past because so many other illnesses have (symptoms of) depression and fatigue and feeling bad in the late afternoon, and too many doctors didn't pay attention. Deep in their hearts they just didn't believe the problem existed and thought they were simply dealing with depressed or nervous patients and had the attitude of "go away and let me deal with real disease."[1]

With hypoglycemia deemed worthy of consideration by the medical profession, perhaps some scientific research will now be done to still the controversy and bring hope to countless sufferers. Although studies have been done over the years, many of them can hardly be regarded as scientific. Science is "knowledge in which the results of investigation have been logically arranged and systematized."[2]

Past research has been done in an arbitrary fashion, not in a logical manner. Most researchers since Whipple's Triad in 1952 have worked with an unscientific premise that the blood sugar must fall to a certain level before hypoglycemia can be diagnosed. Does this invalidate the research of Reaven at Stanford, of Permutt in St. Louis, Lefebvre in Belgium, Hofeldt in Denver, or Fajans and Floyd in Ann Arbor? Each used arbitrary numbers in his investigations.

Reaven says the blood sugar must fall to 40 before he considers hypoglycemia.[3] LeFebvre uses the figure 45,[4] while Hofeldt advocates 40.[5] Fajans and Floyd determine 40 to 45[6] as the area to watch, and Permutt employs 50.[7] Most others also treat 50 as the cut-off point. Are any of them right? Was the research scientific? The glucose tolerance test has been the only means for detecting hypoglycemia. Many physicians realize, however, that this test should be replaced. But what should the new method be?

Before the answers to these and other questions can be determined, the sciences of medicine and nutrition must get together. The conflict between the two must be resolved. United, "nutri-med" can freely and conclusively decide what constitutes this metabolic dysfunction and how it can better be diagnosed.

Realizing several years ago that more research needs to be accomplished, I dreamed of establishing an organization which would help to resolve some of these problems. The result became RIME – Research Into Metabolic Errors. Today, no longer a dream, RIME is a non-profit research organization in the State of California.

Many authorities believe that hypoglycemia is responsible for serious problems in today's society, but not enough research has been done to substantiate this theory. RIME's program is to draw together top-ranking nutritionists and nutrition-oriented physicians to do carefully controlled, scientific studies. Unproved, un-scientific conclusions will either be proven or discarded and new scientific data will be obtained and dispersed. Several areas have been targeted for such research, including schools, jails, prisons, mental institutions, and alcoholic rehabilitation centers.

Statistical information has clearly shown that children are desperately in need of immediate help. Twenty-five percent of children in the United States have learning problems which, many professionals believe, can be attributed to sugar-starved brains. Seventy-six percent of a group of 265 hyperkinetic children were found to be hypoglycemic in a recent study done in New York.[8]

If blood sugar problems can be detected and controlled in children, perhaps there will be less need for jails, prisons, mental institutions, alcoholic rehabilitation centers and the like. If hypoglycemia can lead to heart trouble and diabetes, as some authorities believe, then controlling it in early life could stamp out the number one and number three killers in our country.

Early detection and control of blood sugar problems would mean educating young people to the value of good nutrition. This is not an easy task, for presently they are acquiring more knowledge of junk food than good food. Television commercials on Saturday mornings make this quite clear.

Fortunately, parents are becoming more aware of what happens to their children after they have eaten food loaded with chemical additives and sugar. Many parents of small children have admitted that after their children have consumed sugar, they become "little monsters," so gradually, they are taking that mood-changing substance away from them. Perhaps eventually, TV commercials acclaiming that sugary cereals and candy are nutritious will someday be removed from them also.

In the meantime, what can consumers do to rid food of unnecessary and harmful substances and to insure better nutrition?

First, *we can read labels carefully.* By allowing more time for grocery shopping, we can find out what's in the products we buy. Everything that contains unbleached or enriched flour, or sugar should be avoided. Whether we have a sugar imbalance or not, good rules to follow are, "If you can't pronounce it, don't eat it" and "If man made it, don't consume it."

Second, *we can learn to recognize the names of additives.* Most chemicals in food and drink are foreign to our bodies and are used as preservatives, buffers, neutralizers, moisture controls, coloring or bleaching agents, or other processing aids. Most of them are put there for the processor's financial benefit, not ours, and therefore, few add nutritional value. So, we must learn to know the difference. Ruth Winter's *A Consumer's Dictionary of Food Additives* is a helpful resource, describing approximately 1,500 of the most often used additives.

Third, we can *write to manufacturers,* letting them know how we feel about all those chemicals, including sugar, going into our food. Contrary to popular belief, consumers do not go unheeded. Several years ago, efforts of concerned mothers prompted baby food manufacturers to remove the sugar and salt from their products.

Sugar can be a prime focus of our attention. The FDA considers sugar a food. But is it? Food is defined as a "nutritive material taken into an organism for growth ... or repair, and for maintaining the vital processes." It has been clearly shown that sugar does none of these things. Scientists have proven that all the nutritive value has been taken from natural sugar in the refinding process and that what remains is a pure chemical—$C_{12}H_{22}O_{11}$.

Dr. Richard L. Hall, vice president for research and development for McCormick and Company, is one of the many authorities contending that sugar is "not ordinarily found in food as it grows. It has to be added. It is, therefore, an additive in every sense of the word."[9]

As long as the FDA considers sugar a food, it can be used indiscriminately by the food processors, and breakfast "foods" will continue to be more like candy than cereal. But if sugar is classified as an additive, its use can be controlled, not only in cereals, but in every processed food. Fructose, dextrose, corn sweetener, corn syrup, and molasses are also forms of sugar, and their use should be controlled as well.

Fourth, *lawmakers need to hear complaints about food additives other than sugar.* Laws pertaining to additives need to be changed for the sake of every American, but especially for those who have difficulty with sugar metabolism. Often, the FDA's hands are tied because of antiquated laws. Only Congress can change them. If certain ones aren't changed, food manufacturers will continue to use tons of chemicals in our foods.

As we have seen, hypoglycemia is a physical condition which sometimes manifests itself in erratic behavior and can be suspected when certain symptoms are present. Once these are recognized, we must take responsibility for our lives and have the problem properly diagnosed and treated.

The following is a list of comprehensive indicators for hypoglycemia. Place a check beside each symptom that applies to you. If no *other* cause has been found for them, you can reasonably suspect hypoglycemia.

Checklist of Hypoglycemia Symptoms

- ☐ Tired for no reason
- ☐ Nervousness
- ☐ Anxieties
- ☐ Weakness
- ☐ Drowsiness
- ☐ Low blood pressure
- ☐ Mental deficiency
- ☐ Worrying unnecessarily
- ☐ Tremors (inside or out)
- ☐ Heart palpitations (heart beats too fast)
- ☐ Mental confusion
- ☐ Hungry all the time
- ☐ Lack of concentration
- ☐ Epilepsy
- ☐ Insomnia
- ☐ Crying spells
- ☐ Convulsions
- ☐ Sweating
- ☐ Alcoholism
- ☐ Allergies
- ☐ Sleepiness
- ☐ Personality changes, schizophrenia or mental illness
- ☐ Anti-social behavior
- ☐ Fears and phobias
- ☐ Fainting or blackouts
- ☐ Poor memory
- ☐ Loss of sex drive (frigidity in women, impotence in men)
- ☐ Muscle pain, (twitching, or spasms)
- ☐ Cold, clammy skin, or cold sweats
- ☐ Obesity
- ☐ Psychosis or neurosis
- ☐ Restlessness
- ☐ Wide mood swings
- ☐ Craving sweets
- ☐ Irritability
- ☐ Speech difficulties
- ☐ Juvenile delinquency
- ☐ Sleep disturbances, (especially sleeping

- few hours, waking up, and can't go back to sleep)
- ☐ Poor appetite
- ☐ Gas on the stomach
- ☐ Stomach burning
- ☐ Sudden increase or decrease in weight
- ☐ Angina
- ☐ Dimness of vision, blurred or double vision
- ☐ Disorientation
- ☐ Unconsciousness
- ☐ High blood pressure
- ☐ Frequent tonsillitis in children
- ☐ Car sickness
- ☐ Hyperactivity
- ☐ Slow learning
- ☐ Headaches
- ☐ Dizziness
- ☐ Constipation
- ☐ Narcolepsy
- ☐ Cold hands and feet
- ☐ Addiction (or craving for) sugar, caffeine, soft drinks, cigarettes, drugs)
- ☐ Incoordination
- ☐ Suicidal tendencies
- ☐ Arthritic pains or arthritis
- ☐ Ulcers (peptic or duodenal)
- ☐ Digestive disorders
- ☐ Staggering (loss of muscle control)
- ☐ Numbness
- ☐ Asthma
- ☐ Gasping for breath
- ☐ Nightmares or night terrors
- ☐ Depression

Today, doctors who use the nutritional approach to medicine and who actively diagnose and treat hypoglycemia can be found in every part of the country. Organizations whose members are such physicians are listed below. A roster of their members are available by writing each group:

Academy of Orthomolecular Psychiatry
1691 Northern Blvd.
Manhasset, New York 11031

American Academy of Medical Preventics
6151 W. Century Blvd., Suite 1114
Los Angeles, California 90045

Huxley Institute
1114 First Ave.
New York, New York 10021

*International Academy of
 Biological Medicine (IABM)*
P. O. Box 31313
Phoenix, Arizona 85046
(Please send stamped, self-addressed No. 10 envelope)

International Academy of Metabology
P. O. Box 15157
Las Cruces, New Mexico 88001

International Academy of Preventive Medicine
10950 Grandview, Suite 469
Overland Park, Kansas 66210
(Enclose $4.00 for postage and handling)

International College of Applied Nutrition
P. O. Box 386
LaHabra, California 90631
(Send stamped, self-addressed envelope)

Orthomolecular Society
P. O. Box 7
Agoura, California 91301

Doctor-members of these organizations are especially mindful of the fact that proper nutrition can maintain the metabolic balance necessary for good health. If they determine that a patient has mental and emotional problems, they will not assign him to a psychiatrist's couch without giving him the proper metabolic tests.

Dr. Sydney Walker III, a neuropsychiatrist from La

Jolla, California, suitably summarizes the declarations made in this book:

> You should not accept the verdict "it's all in your head" if your doctor has neglected to give you a thorough examination, especially if you have physical symptoms in addition to emotional ones. Too many people have been called emotionally disturbed when physical factors were contributing to their condition.
>
> Because carbohydrate problems are so subtle, it makes little sense to take a position "pro" or "con" on hypoglycemia or to try to turn it into a social issue. Each case must be evaluated without bias. And that's the unsugar-coated truth.[10]

That statement does not imply that hypoglycemics never need the assistance of a psychiatrist. When persons have had low blood sugar and its often-accompanying emotional disturbances for a long time, they may need to continue to see a psychiatrist, even though the hypoglycemia has been brought under control. Their habits may have become so ingrained that professional help is still needed.

But in the final analysis, the individual is responsible for seeking out a doctor who "believes" in hypoglycemia. If no trace of a blood sugar problem can be found after the proper tests, then another solution for the discomfort must be sought. And if the blood sugar is abnormally low, eating habits must change to get the level back to normal.

Many things, including crime, can be precipitated by poor nutrition. Endless numbers of graduates from medical schools have learned what to do when our bodies go into crisis with disease, broken bones, and other injuries. But few of them are taught how to aid and encourage the nourishment and growth of those bodies – sometimes called preventive medicine.

Principles of nutrition can be studied and even learned, but until they become a way of life, the joy of continuous good health is never experienced. It is only when we see the positive effects that pure food has on our physical, emotional and spiritual well-being that we can sound the alarm to others that *Sugar Isn't Always Sweet!*

Appendix A/Articles on Childhood Hypoglycemia

Beesley, C. A. "Neonatal Hypoglycemia – a Potential Cause of Damage?" *Australian Nursing Journal* 19(1980):45–46.

Cornblath, M.; Joassin, G.; Weiskopf, B.; and Swiatek, K. R. "Hypoglycemia in the Newborn." *Pediatrics Clinics of North America* 13(1966):905–20.

Greenberg, Robert E. and Christiansen, Robert O. "The Critically Ill Child: Hypoglycemia." *Pediatrics* 46(1970):915–20.

Hirschhorn, N.; Lindenbaum, J.; Greenough, W. B. III; and Alam, S.; "Hypoglycemia in Children with Acute Diarrhea." *Lancet* 2(1966):128–132.

"Hypoglycemia in the Newborn." *Medical Journal of Australia* 2(1970):4.

Kelnar, C. J. "Hypoglycemia in Children Undergoing Adenotonsillectomy." *British Medical Journal* 1(1976):751–52.

Maze, A., and Samuels, S. I. "Hypoglycemia-induced Seizures in an Infant During Anesthesia." *Anesthesiology* 52(1980):77–78.

Milner, R. D. "Neonatal Hypoglycemia – a Critical Reappraisal." *Archives of Diseases of Children* 47(1972):679–82.

Mistra, P. K., and Sharma, B. "Hypoglycemia in Newborns – a Prospective Study." *Indian Pediatrics* 14(1977):129–32.

Molla, A. M.; Hossain, M.; Islam, R.; Bardhan, P. K.; and Sarker, S. A. "Hypoglycemia: a Complication of Diarrhea in Childhood." *Indian Pediatrics* 18(1981):181–85.

Pagliara, D. S.; Karl, I. E.; Haymond, M.; and Kipnis, D.

M. "Hypoglycemia in Infancy and Childhood." *Journal of Pediatrics* 82(1973):365-79.

Pak, Yong Cha; Limbeck, George A.; and Kelley, Vincent C. "Hypoglycemia in Children – Relationship Between Adrenocortical Function and Growth." *American Journal of Diseases of Children* 12(1971):10-14.

Stevens, D. and Schreiner, R. L. "Neonatal Hypoglycemia." *Journal of the Indiana State Medical Association* 71(1978):458-67.

Ware, S., and Osborne, J. P. "Postoperative Hypoglycemia in Small Children." *British Medical Journal* 2(1976):499-501.

Wolf, R. L. "Hypoglycemia – an Important Neonatal Problem." *South African Nursing Journal* 42(1975):17-19.

Wood, C. B., and Mann, T. P. "Hypoglycemia in Childhood." *Postgraduate Medical Journal* 42(1966):555-61.

Appendix B/Less Frequently Named Symptoms of Hypoglycemia

Negativism, apathy, *emotional instability,* sudden loss or gain of weight, *absentmindedness,* lightheadedness, *indecisiveness,* stomach pains, abdominal spasms, loss of appetite, lapse of moral conduct, *hostility* and beligerance, *lack of energy,* listlessness, excitability, hypochondria, hypothermia (low temperature), hearing problems, multiple sclerosis, coma, ringing in the ears, sighing and yawning, sensitivity to noise and/or light, itchy skin, smothering spells, *vague sense of dread,* temper tantrums, underachievement in school, uncontrollable temper, *shrewishness,* eczema, maniacal behavior, acute paranoia, nervous breakdown, boredom, nausea, flatulence (gas), amnesia, motionsickness, burning stomach, angina pains, disorientation, double vision, *agitation,* antagonism, violence, lying, hallucinations, pallor, dimness of vision, flickering before the eyes, increased flow of saliva, painful discharge of urine, difficulty in forming words, inequality of pupils, torsion (twitching or wrenching of the body), grimacing, gesticulation, querlouessness, bizarre behavior, slowness of thought, inclination to loiter and dawdle or give random answers, compulsions, impulsive actions, wandering and stupor (Symptoms in italics are the ones I had [Author]).

Not all those who write about hypoglycemia mention symptoms, and those who do mention only a few. This list is a compilation of symptoms from all the books, magazines and medical journals surveyed.

Appendix C/Medical Schools with Required Courses in Nutrition

Of 129 medical schools in the United States, only twelve have required courses in nutrition. Information obtained from *AAMC Curriculum Directory* (Washington, D.C.: Association of American Medical Colleges, 1979-80).

Emory University School of Medicine
Atlanta, GEORGIA 30322

School of Medicine at Morehouse College
223 Chestnut Street, S.W.
Atlanta, GEORGIA 30314

University of Hawaii, John A. Burns School of Medicine
1960 East-West Road
Honolulu, HAWAII 96822

University of Rochester
601 Elmwood Avenue
Rochester, NEW YORK 14642

Columbia University College of Physicians and Surgeons
630 W. 168th Street
New York, NEW YORK 10032

State University of New York, Upstate Medical Center
766 Irving Avenue
Syracuse, NEW YORK 13210

University of North Carolina School of Medicine
Chapel Hill, NORTH CAROLINA 27514

Northeastern Ohio University College of Medicine
Rootstown, OHIO 44272

University of Oregon Health Sciences Center
School of Medicine

3181 S. W. Sam Jackson Park Road
Portland, OREGON 97201

Medical College of Pennsylvania
3300 Henry Avenue
Philadelphia, PENNSYLVANIA 19129

Brown University Division of Biology and Medicine
97 Waterman Street
Providence, RHODE ISLAND 02912

University of Tennessee College of Medicine
800 Madison Avenue
Memphis, TENNESSEE 38163

Appendix D/
Currier Pancakes

1 cup 7-grain granola
⅓ cup wheat germ untoasted
   (only if fresh dated is available)
4 to 6 tbsp. lecithin granules
1 to 1½ cups Stone Buhr 100%
   whole wheat pancake mix or
   Aunt Jemima 100% whole wheat mix
1 level tbsp. bone meal
⅓ cup bran flakes
4 tbsp. Brewer's yeast or Torulo yeast

Mix dry ingredients in large bowl. In a blender, mix 2 cups non-fat milk (for more protein, non-fat powdered milk can be added), 2 or 3 fertile eggs. Blend into dry ingredients.

For flavor treats, add 1 cup unsweetened apple sauce, or 1 unpeeled apple with cores and seeds removed, banana, fresh pineapple or strawberries. Use your good nutrition imagination!

Appendix E/
Currier Cocktail

- 1 or 2 glasses non-fat milk, certified raw (if possible)
- 1 egg, fertile (if available)
- 2 tbsp. non-fat powdered milk
- 3 tbsp. wheat germ and/or rice polishings (wheat germ only if fresh dated is available)
- 1 tbsp. natural bran flakes
- 1 tbsp. primary food yeast or Brewer's yeast
- 1 tbsp. lecithin granules

This drink may be flavored with the following: plain yogurt (commercial fruit yogurt has sugar), banana, natural applesauce, carob powder, powdered coconut, or unsweetened fruit juices. Always keep in mind that no sugar or anything containing sugar may be added. Use blender if possible. Keep ingredients in refrigerator. This egg nog may be flavored to your own preference.

Notes

Chapter 4   No One Is Immune

[1] Diana Bovill, "A Case of Functional Hypoglycemia – a Medico-Legal Problem," *British Journal of Psychiatry* 123 (1973): 353-58.

[2] Carlton Fredericks, *Low Blood Sugar and You* (New York: Gosset & Dunlap, 1969).

Chapter 5   Where Are You, God?

[1] Psalm 42:5, 9 (RSV).

[2] Psalm 43:1, 2 (RSV).

[3] George Watson, *Nutrition and Your Mind* (New York: Harper & Row, 1972), p. 21.

Chapter 6   Stepchild of Medicine

[1] *Hypoglycemia and Me* (Troy, New York: Adrenal Research Society of the Hypoglycemia Foundation Inc., n.d.).

[2] A.E. Bender, "Some Metabolic Effects on Dietary Sucrose," *Nutrition and Metabolism*, December 1970, pp. 22-39.

[3] Jerrold M. Milstein, "Hypoglycemia in the Neonate," *Postgraduate Medicine* 50 (1971): 91-94.

[4] "Hypoglycemia in the New-born," *Medical Journal of Australia*, 2 (1970): 4.

[5] Seale Harris, "Hyperinsulinism and Dysinsulinism,"

Journal of the American Medical Association 83 (1924): 729-33.

6 Paavo Airola, *Hypoglycemia: a Better Approach*, 9th ed. (Phoenix: Health Plus Publishers, 1977), p. 80.

7 "Sweet Tooth, Sour Facts," *Time*, January 13, 1958, p. 34.

8 "Sugar, Can It Be a Health Hazard?" *Good Housekeeping*, August 1973, p. 135.

9 John Feltman, "You Can Live Without Sugar," *Organic Gardening*, October 1974, p. 90.

10 Raymond Sokolov, "The Most Artificial Sweetener of Them All," *Natural History*, April 1975, pp. 86-90.

11 John Pekkanen, "Sweet and Sour," *Atlantic Monthly*, July 1975, pp. 50-53.

12 "Too Much Sugar?" *Consumer Reports*, March 1978, pp. 136-42.

13 "Sugar, Enemy of Good Nutrition," *Consumer Bulletin*, February 1961, pp. 10-31.

14 "Sugars, Coronaries Linked," *Science Newsletter*, August 29, 1964, p. 31.

15 Gerald M. Knox, ed., "Sugar, Those Silent Sugar Diseases: Some Facts You Should Know," *Better Homes and Gardens*, February 1973, pp. 30-32.

16 Jane Ogle, "Killer on the Breakfast Table," *Harper's Bazaar*, September 1973, p. 150.

17 Jane E. Brody, "Sugar, Villain in Disguise," *Reader's Digest*, October 1977, pp. 163-65.

18 Chris W. Lecos, "Sugar, How Sweet It Is—and Isn't," *FDA Consumer*, February 1980, pp. 21-23.

19 "Sugar, No. 1 Murderer," *Nutrition News*, 1978, p. 1.

20 "Hidden Sugars in Food," Kurt W. Donsbach, Huntington Beach, California, n.d.

21 Ross Hume Hall, *Food for Nought, the Decline in Nutri-*

tion (Hagerstown, Maryland: Harper & Row, 1974), pp. 253-54.

[22] Alexander G. Schauss, *Orthomolecular Treatment of Criminal Offenders* (Berkeley: Michael Lesser, M.D.), 1978.

Chapter 7   Millions Suffer Needlessly

[1] Seale Harris, "Hyperinsulinism, a Definite Disease Entity," *Journal of American Medical Association* 101 (1933): 1958-65.

[2] Sidney A. Portis, "Life Situation, Emotions and Hyperinsulinism," *Journal of American Medical Association* 101 (1950): 1281-86.

[3] *AAMC Curriculum Directory* (Washington, D.C.: Association of American Medical Colleges, 1982).

[4] Harris, "Hyperinsulinism."

[5] Holbrooke S. Seltzer, "Spontaneous Hypoglycemia as an Early Manifestation of Diabetes Mellitus," *Diabetes* 5 (1956): 437-42.

[6] Robert D. Gittler, "Spontaneous Hypoglycemia," *New York State Journal of Medicine* 62 (1962): 236.

[7] Albert J. Finestone, "Hypoglycemia in the Fed State," *Medical Clinics of North America* 54 (1970): 540.

[8] Paul T. Chandler, "An Update on Reactive Hypoglycemia," *American Family Physician* 16 (1977): 113.

[9] Robert C. Atkins' introduction to Ruth Adams' *Is Low Blood Sugar Making You a Nutritional Cripple?* (New York: Larchmont Books, 1975), pp. 6, 7.

[10] Ibid., p. 8.

[11] Emmett B. Carmichael, "Seale Harris—Physician-Scientist-Author," *Journal of Medical Association of Alabama* 33 (1964): 189-95.

[12] P.J. Cammidge, "Chronic Hypoglycemia," *The Practi-*

    tioner 119 (1927): 102-12.
13. James Greenwood Jr., "Hypoglycemia as a Cause of Mental Symptoms," *Pennsylvania Medical Journal* 39 (1935): 12-16.
14. Ibid.
15. A.R. Vonderahe, "Personality Change in Hypoglycemia," *Journal of Medicine* 17 (1936): 189-90.
16. Joseph Wilder, "Psychological Problems in Hypoglycemia," *American Journal of Digestive Diseases* 10 (1941): 428-35.
17. Margaret Bass, "Psychiatric Aspects of Spontaneous Hypoglycemia," *Journal of Nervous and Mental Disease* 108 (1948): 7-8.
18. Richard H. Hoffman, "Hyperinsulinism, a Factor in Neuroses," *American Journal of Digestive Disease* 16 (1949): 242.
19. Sidney A. Portis, "Life Situation, Emotions and Hyperinsulinism," *Journal of American Medical Association* 142 (1950): 1281-86.
20. Wilder, "Problems of Criminal Psychology Related to Hypoglycemic States," *Journal of Criminal Psychology* 1 (1940): 219-33.
21. David Adlersberg, "Medico-Legal Problems of Hypoglycemic Reactions in Diabetes," *Annals of Internal Medicine* 12: 1804-15, 1939.
22. Harry M. Salzer, "Relative Hypoglycemia as a Cause of Neuropsychiatric Illness," *Journal of the National Medical Association* 58 (1966): 12-17.
23. "Statement on Orthomolecular Medicine," *Orthomolecular Medical Society's Referral Booklet*, San Francisco, n.d.
24. "Statement on Hypoglycemia," *Journal of American Medical Association* 223 (1973): 682; *Annals of Internal Medicine* 78 (1973): 300-01; *Archives of Internal Medicine* 131 (1973): 591; *Diabetes* 22 (1973): 137.

[25] Wilder, "Problems of Criminal Psychology," pp. 220, 228.

[26] Wilder, "Psychological Problems in Hypoglycemia," p. 34.

Chapter 8 Discovering the Enemy

[1] Allen O. Whipple, "Islet Cell Tumors of the Pancreas," *Canadian Medical Association Journal* 66 (1952): 334–41.

[2] Allan L. Drash, "Taking the Guesswork Out of Hypoglycemia," *Patient Care*, April 15, 1972, p. 20.

Chapter 9 Eat Right and Save Your Life

[1] William Dufty, *Sugar Blues* (New York: Warner Books, 1975), p. 82.

[2] Ibid., p. 78.

Chapter 10 The Hassles in Getting Good Food

[1] *1958 Food Additives Amendment to the Federal Food, Drug and Cosmetic Act*, 1980, p. 4.

[2] Phyllis Lehmann, "More Than You Ever Thought You Would Know About Food Additives ... Part I," *FDA Consumer*, April 1979, p. 10.

[3] Michael Walczak and Harold S. Stone, eds., *Nutrition Applied Personally* (LaHabra, California: International College of Applied Nutrition, revised 1981).

[4] Michael Jacobson, *Eater's Digest* (Garden City, New York: Doubleday & Company, 1972), p. 9.

[5] *Food Processing*, June 1982.

[6] Ibid.

[7] Ibid.

[8] Ibid.

[9] Ruth Winter, *A Consumer's Dictionary of Food Additives* (New York: Crown Publishers Inc., 1978), p. 163.

10. Ross Hume Hall, *Food For Naught, the Decline in Nutrition* (Haggerstown, Maryland: Harper & Row, 1974), pp. 61, 62.
11. *Public Law* 95-203, 95th Congress, November 23, 1977.
12. David Reuben, *Everything You Always Wanted to Know About Nutrition* (New York: Avon Books, 1979), p. 190.
13. Winter, *Consumer's Dictionary*, p. 207.
14. Lehmann, "More Than ... Part III," *FDA Consumer*, June 1979, p. 13.
15. Winter, *Consumer's Dictionary*, p. 112.
16. Message from the President of the United States, 91st Congress, 1st Session, House of Representatives Document No. 91-188.
17. Lehmann, "More Than ... Part I," *FDA Consumer*, April 1979, p. 12.

Chapter 11   The Rest of the Deadly Don'ts

1. "Caffeine Caution," *Sciequest*, September 1980, p. 17.
2. Paavo Airola, *Hypoglycemia, a Better Approach* (Phoenix: Health Plus Publishers, 1977), p. 60.
3. Matt Clark, "Is Caffeine Bad for You?" *Newsweek*, July 19, 1982, p. 63.
4. Ibid., pp. 62-64.
5. *Department of Agriculture* figures for October 1982.
6. Walter S. Ross, "What's Been Added to Your Cigarettes?" *Reader's Digest*, July 1982, pp. 111-14.
7. Robert C. Atkins, *Diet Revolution* (New York: David McKay Company, 1972), p. 273.
8. E. Cheraskin, *Psycho-Dietetics* (New York: Bantam Books, 1974), pp. 44, 45.
9. Airola, *Hypoglycemia: a Better Approach*, p. 62.
10. Roy C. Gumpel, "Alcohol-Induced Hypoglycemia," *New*

York State Journal of Medicine 64 (1964): 1014-17.

[11] "Alcoholic Hypoglycemia," *Nutrition Reviews* 23 (1965): 43-45.

[12] Harry L. Gibbons, "Alcohol-Induced Hypoglycemia as a Factor in Aircraft Accidents," *Aerospace Medicine* 37 (1966): 959-61.

Chapter 12   RIME, Reason and Rescue

[1] " 'Health-fad' Illness Gets Respect," *Washington Post*, Washington, D.C., December 28, 1982.

[2] *Funk & Wagnalls Standard Desk Dictionary.*

[3] Gerald M. Reaven, *personal interview*, Stanford University, Palo Alto, California.

[4] P.J. Lefebvre, "Studies on the Pathogenesis of Reactive Hypoglycemia: Role of Insulin and Glucagon," *Hormone and Metabolic Research*, Rome: 1976 European Symposium.

[5] Fred D. Hofeldt, "Diagnosis and Classification of Reactive Hypoglycemia," *American Journal of Nutrition* 25 (1972): 1193-1201.

[6] Stefan S. Fajans, "Hypoglycemia: How to Manage a Complex Disease," *Modern Medicine* 2 (1973): 24-31.

[7] M.A. Permutt, "Postprandial Hypoglycemia," *Diabetes* 25 (1976): 719-33.

[8] J. Langseth, "Glucose Tolerance and Hyperkinesis," Great Britain: Pergammon Press, *Federal Cosmetic Toxicology* 16 (1977): 129-33.

[9] Richard L. Hall, "Food Additives," *Nutrition Today*, July-August 1973, pp. 20-28.

[10] Sydney Walker III, "Sugar Doctors Rush Hypoglycemia," *Psychology Today*, July 1975, pp. 68-70.

Bibliography/Books

Abrahamson, E.M., and Pezet, A.W. *Body, Mind and Sugar.* New York: Holt, Rinehart & Winston, 1951.

Adams, Ruth, and Murray, Frank. *Is Low Blood Sugar Making You an Emotional Cripple?* New York: Larchmont Books, 1970.

Airola, Paavo. *Hypoglycemia, a Better Approach.* Phoenix: Health Plus Publishers, 1977.

Atkins, Robert H. *Diet Revolution.* New York: David Mckay Company Inc., 1972.

Barmakian, Richard. *Hypoglycemia, Your Bondage or Your Freedom.* Irvine, California: Altura Health Publishers, 1976.

Blaine, Tom R. *Goodbye Allergies.* Secaucus, New Jersey: Citadel Press, 1965.

Brennan, R.O. *Nutrigenetics, New Concepts for Relieving Hypoglycemia.* Denver: Nutri-Books Corporation, 1975.

Cheraskin, Emanuel; Ringsdorf, W.M. Jr.; Clark, J.W. *Diet and Disease.* Emmaus, Pennsylvania: Rodale Press Inc., 1968.

Cheraskin and Ringsdorf. *New Hope for Incurable Disease.* New York: Exposition Press, 1971.

Cheraskin, Ringsdorf and Brecher, A. *Psycho-Dietetics.* New York: Bantam Books, 1976.

Crook, Wm.G. *Are You Bothered by "Hypoglycemia."* Johnson, Tennessee: Professional Books, 1977.

Dinaburg, Cathy, and Abel, D'Ann. *Nutrition Survival Kit.* San Francisco: Panjandrum Press and Aris Books, 1976.

Donsbach, Kurt. *Hypoglycemia.* Huntington Beach, California: International Institute of Natural Health, 1976.

Dufty, William. *Sugar Blues.* New York: Warner Books Inc., 1976.

Fredericks, Carlton, and Goodman, Herman. *Low Blood Sugar and You.* New York: Grossett & Dunlap, 1969.

Greenblatt, Robert B. *Search the Scriptures.* New York: J. B. Lippincott Company, 1963.

Hall, Ross Hume. *Food For Naught.* Hagerstown, Maryland: Harper & Row, 1974.

Hunter, Beatrice Trum. *Great Nutrition Robbery.* New York: Charles Scribner's Sons, 1978.

Hunter. *The Sugar Trap.* Boston: Houghton, Mifflin Company, 1982.

Hurdle, J. Frank. *Low Blood Sugar.* Newberry Park, Massachusetts: Parker Publishing Company, 1969.

Hypoglycemia and Me? Troy, New York: Adrenal Metabolic Research Society of the Hypoglycemia Foundation, n.d.

Jacobson, Michael F. *Eater's Digest.* New York: Doubleday & Company, 1972.

Martin, Clement. *Low Blood Sugar, Hidden Menace of.* New York: Arco Publishing Company Inc., 1970.

Nittler, Allan. *New Breed of Doctor.* New York: Bantam Books, 1974.

Reuben, David. *Everything You Always Wanted to Know About Nutrition.* New York: Avon Books, 1978.

Roberts, H.J. *The Causes, Ecology and Prevention of Traffic Accidents.* Springfield, Illinois: Charles C. Thomas, 1971.

Roberts, Samuel. *Exhaustion.* Emmaus, Pennsylvania: Rodale Press Inc., 1967.

Ross, Harvey. *Fighting Depression.* New York: Larchmont Books, 1975.

Saunders, Jeraldine, and Ross, Harvey. *The Disease Your Doctor Won't Treat.* New York: Pinnacle Press, 1980.

Steinchron, Peter J. *Low Blood Sugar.* New York: New American Library, 1972.

Thosteson, George. *Help for Hypoglycemia.* Chicago: Field Enterprises Inc., 1974.

Watson, George. *Nutrition and Your Mind.* New York: Harper & Row, 1972.

Weller, Charles. *How to Live with Hypoglycemia.* New York: Doubleday & Company, 1968.

Winter, Ruth. *A Consumer's Dictionary of Food Additives.* New York: Crown Publishers Inc., 1978.

Bibliography/Textbooks

Bauer, John D., et al. *Clinical Laboratory Methods.* St. Louis: Mosby, C.V. Co., 1974.

Beeson, Paul B., ed. *Cecil-Loeb Textbook of Medicine.* Philadelphia: W. B. Saunders Company, 1979.

Bolinger, Robert E., ed. *Endocrinology.* Garden City, New York: Medical Examination Publishing Company, 1978.

Conn, Howard F., and Conn, Rex B. *Current Diagnosis.* Philadelphia: W. B. Saunders Company, 1977.

Hart, F. Dudley, ed. *French's Index of Differential Diagnosis.* Chicago: Year Book of Medicine, 1979.

Harvey, A. McGhee, et al., eds. *Differential Diagnosis.* Philadelphia: W. B. Saunders Company, 1979.

Isselbacher, Kurt J., et al. *Harrison's Principles of Internal Medicine.* New York: McGraw, 1980.

Krupp, Marcus A., ed. *Current Medical Diagnosis & Treatment.* Los Altos, California: Lange Medical Publications, 1980.

Krupp, Marcus A., ed. *Physician's Handbook.* Los Altos, California: Lange Publishing Company, 1976.

Nichols, Armand M. Jr. ed. *Harvard Guide to Modern Psychiatry.* Cambridge, Massachusetts: Harvard University Press, 1978.

Salkin, Paul, ed. *Psychiatry.* Garden City, New York: Medical Examination Publishing Company Inc., 1977.

Taylor, R.B., et al., eds. *Family Medicine, Principles and Practice.* New York: Springer-Verlag, 1978.

Vaugnn, Victor C., ed. Nelson. *Textbook of Pediatrics.* Philadelphia: W. B. Saunders Company, 1980.

Williams, Robert H., ed. *Textbook of Endocrinology.* Oxford, England: Oxford University Press, 1980.

Bibliography/Periodicals

Arehart-Treichel, Joan. "The Great Medical Debate Over Low Blood Sugar," *Science News*, March 17, 1973, pp. 172-74.

Cherry, Laurence. "The Bitter News About Sugar and Salt." *Glamour*, March 1979, p. 246.

Cole, William. "Hypoglycemia, Shortage of Body Fuel." *Today's Health*, November 1968, pp. 40-43.

"Coping with Hypoglycemia." *Awake!*, July 22, 1978, pp. 5-8.

Dunn, Roz. "Sugar: Kicking the Habit." *Essence*, January 1978, p. 54.

Endicott, Wendy. "Sweet Poison." *Harper's Bazaar*, October 1969, p. 5.

Findley, Paul, U.S. Representative from Illinois. "Sugar, a Sticky Mess in Congress." *Reader's Digest*, June 1966, pp. 88-91.

Fishbein, Morris. "Hypoglycemia." *Consumer Bulletin*, October 1972, p. 18.

Frank, Arthur, and Frank, Stuart. "Hypoglycemia." *Madamoiselle*, January 1975, pp. 134-45.

"He Solves Psychiatric Problems with Nutrition." *Prevention*, June 1972, pp. 69-78.

Hess, John L. "Harvard's Sugar-Pushing Nutritionist." *Saturday Review*, August 1978, p. 10.

Hopkins, Harold. "The GRAS list Revisited." *FDA Consumer*, May 1978, pp. 13-15.

"Is Low Blood Sugar *Really* a Major Health Problem?" *Good Housekeeping*, May 1972, p. 213.

"Is Sweetness Your Weakness?" *Essence*, August 1979, p. 66.

Johnson, D. Kendrick. "Another Man's Poison." *IN*, November 1978, pp. 31, 32.

Keough, Carol. "Rating Those Food Additives." *Organic Gardening*, February 1980, pp. 166-70.

Kolata, Gina Bari, "The Truth About Hypoglycemia." *Science Magazine*, November 1979, pp. 26-30.

Kushner, Trucia. "Hypoglycemia." *Madamoiselle*, April 1975, p. 141.

Lamberg, Lynne. "Sweet Deception." *Working Woman*, December 1980, p. 98.

Liebman, Bonnie F., and Moyer, Greg. "The Case Against Sugar." *Nutrition Action*, December 1980, p. 9.

"Low Blood Sugar: Fact and Fiction." *Consumer Report*, July 1971, p. 444.

Mayer, Jean. "Sugar: Taking the Bitter with the Sweet." *Family Health*, March 1978, pp. 2626-7.

Mayer. "The Bitter Truth About Sugar." *New York Times Magazine*, July 30, 1976, p. 26.

Meislin, Rich. "Sugar Increases Blood Fats." *Prevention*, December 1972, pp. 160-65.

Miller, Roger. "In Science We Trust (Unless It Involves Sugarless Soda)." *FDA Consumer*, May 1981, p. 27.

Nolen, William A. "Low Blood Sugar—What It Means." *McCalls*, November 1975, pp. 92-94.

Noonan, Jacqueline A., and Hart, B.F. "Hypoglycemia and Its Cousins." *Let's Live*, September, 1979, pp. 15-19.

Null, Gary, and Null, Steve. "Sweet Suicide." *Let's Live*, August 1977, pp. 24-29.

O'Connell, Barbara. "Hypoglycemia, the Disease That Makes Women Tired." *Science Digest*, March 1969, pp. 40-44.

Pauling, Linus. "Sugar: Sweet and Dangerous." *Executive Health*, 1972, p. 1.

Siegel, Paula. "The Sweet Truth—How to Beat the Sugar Habit." *Gentlemen's Quarterly*, October 1979, p. 90.

Simon, Nissa. "How Sugar Gets to Your Skin—and Harms It." *Vogue*, May 1977, pp. 108, 109.

Solomon, Neil. "Are You a Sugar Addict?" *Harper's Bazaar*, July 1975, pp. 26, 27.

Stern, Judith, "Sugar Yes? No?" *Vogue*, August 1975, p. 76.

"Sweet Talk About Sweetened Cereals. *Consumer Reports*, March 1980, p. 140.

Switzer, Ellen. "Your Moods and Blood Sugar." *Vogue*, October 1973, p. 226.

"The Facts About 'Low Blood Sugar'." *Changing Times*, September 1973, pp. 13-15.

"The Fad Disease." *Time*, April 7, 1980, p. 71.

"The Hypoglycemia Fad." *Newsweek*, January 29, 1973, p. 47.

"The Truth About Breakfast Cereals." *Let's Live*, September 1970, p. 42.

"Too Much Sugar." *Consumer Reports*, March 1978, pp. 136-39.

Trotter, Robert J. "Agression, a Way of Life for the Quolla." *Science News*, February 3, 1973, pp. 76, 77.

Underwood, Don, and Thurston, Emory W. "Low Blood Sugar." *Research Bulletin, Institute of Nutritional Research*, May 1971, p. 1.

Wade, Carlson. "The Sweet Killer in Your Cupboard." *Natural Food and Farming*, October 1978.

Wagner, Madeline. "Hypoglycemia: the Deceptive Disease." *Cosmopolitan*, November 1980, p. 126.

Webb, Linda. "Exploding the Myths About Sugar." *Good Housekeeping*, October 1979, p. 176.

Webb. "You and Your Diet, an Up-to-Date Report on Low Blood Sugar." *Good Housekeeping*, November 1978, p.

318.

Welch, Mary Scott. "Hypoglycemia." *Ladies' Home Journal,* July 11, 1971, p. 98.

"What's All the Fuss About Low Blood Sugar?" *Changing Times.* December 1959, pp. 15, 16.

"Yesterday's Additives—Generally Safe." *FDA Consumer,* March 1981, pp. 14, 15.

Yuncker, Barbar. "Sugar and the New Theory About Heart Attacks." *House and Garden,* February 1970, pp. 64, 65.

Bibliography/Medical Journals

Abell, D.A. "The Significance of Abnormal Glucose Tolerance (Hyperglycemia and Hypoglycemia) in Pregnancy." *British Journal of Obstetrics and Gynecology* 86 (1979): 214-21.

Adolfsson, Rolf; Bucht, Gösta; Lithner, Folke; and Winblad, Bengt. "Hypoglycemia in Alzheimer's Disease." *Acta Medica Scandinavica* 208 (1980): 387-88.

Andreasen, A.T. and Maraspini, Christiana. "Symptomless Hypoglycemia." *South African Medical Journal* 50 (1976): 1339-41.

Baker, Frank J, II; Rosen, Peter.; Coppleson, Lionel W.; Evans, Tom; Fauman, Beverly; and Segal, Marshall B. "Diabetic Emergencies: Hypoglycemia and Ketoacidosis." *Journal of American Emergency Physicians* 5 (1976): 119-22.

Baker, Robert J. "Newer Considerations in the Diagnosing and Management of Fasting Hypoglycemia." *Surgical Clinics of North America* 49 (1969): 191-206.

Beeuwkes, Adelia M. "The Dietary Treatment of Functional Hyperinsulinism." *Journal of American Dietetic Association* 18 (1942): 731-33.

Berger, Herbert. "Hypoglycemia: a Perspective." *Postgraduate Medicine* 57 (1975): 81-85.

Block, Marshall B. "Hypoglycemia: Clinical Implications, Part I: Reactive Hypoglycemia." *Arizona Medicine* 31: (1974): 932-34.

Buckley, Robert E. "Hypoglycemia, Temporal Lobe Disturbance and Aggressive Behavior." *Orthomolecular Psychiatry* 8 (1979): 188-92.

Burke, M. Desmond. "Hypoglycemia: Test Strategies for Laboratory Evaluation." *Frontiers in Medicine* 6 (1980): 39-42.

Burman, Kenneth D.; Cunningham, Edwin J.; Klachko, David M.; Bazzoui, Widad E.; and Burns, Thomas W. "Factitious Hypoglycemia." *American Journal of the Medical Sciences* 266 (1973): 23-30.

Calbreath, Donald F. "The Glucose Tolerance Test." *Nursing Care*, October 1977, pp. 26, 27.

Cammidge, P.J. "Chronic Hypoglycemia." *British Medical Journal* 1 (1930): 818-22.

Carter, W. Phelps Jr. "Hypothermia – a Sign of Hypoglycemia." *Journal of American College of Emergency Physicians* 5 (1976): 594-95.

Chambers, William H. "Undernutrition and Carbohydrate Metabolism." *Physiological Review* 18 (1938): 248-88.

Chandler, Paul T. "An Update on Reactive Hypoglycemia." *American Family Physician* 16 (1977): 113-519.

Chase, H. Peter; Marlow, Robert A.; Dabiere, Carol S.; and Welch, Noreen. "Hypoglycemia and Brain Development." *Pediatrics* 52 (1973): 513-19.

Cohen, Saul. "A Review of Hypoglycemia and Alcoholism With or Without Liver Disease." *Annals New York Academy of Sciences* 273 (1976): 339-42.

Colt, Edward. "Heart Failure and Hypoglycemia." *New York State Journal of Medicine* 76 (1976): 2033.

Conn, Jerome W. "The Spontaneous Hypoglycemias." *Journal of American Medical Association* 115 (1940): 1669-74.

Conn and Newburgh, L. H. "The Glycemic Response to Isoglucogenic Quantities of Protein and Carbohydrate." *Journal of Clinical Investigation* 15 (1936): 665-71.

Conn and Seltzer. "Spontaneous Hypoglycemia." *American Journal of Medicine* 19 (1955): 460-78.

Danowski, T.S.; Nolan, S.; and Stephan, T. "Hypoglycemia." *World Review of Nutrition and Dietetics* 22 (1975): 288-303.

Drew, John H.; Abell, David A.; and Beischer, Norman A. "Congenital Malformations, Abnormal Glucose Tolerance, and Estriol Excretion in Pregnancy." *Journal of American College of Obstetricians and Gynecologists* 51 (1978): 129-31.

Fabrykant, Maximilian. "The Problem of Functional Hyperinsulinism or Functional Hypoglycemia Attributed in Nervous Causes." *Metabolism* 4 (1955): 469-78.

Fajans, Stefan S. and Floyd, John C. Jr. "Hypoglycemia: How to Manage a Complex Disease." *Modern Medicine*, October 15, 1973.

Fink, William J.; Hucke, Samuel T.; Gray, Thomas W.; Thompson, Bernard W.; and Read, Raymond C. "Treatment of Postoperative Reactive Hypoglycemia by a Reversed Intestinal Segment." *American Journal of Surgery* 131 (1976): 19-22.

Fitzgerald, Faith T. "Hypoglycemia and Accidental Hypothermia in an Alcoholic Population." *Western Journal of Medicine* 133 (1980): 105-107.

Ford, Charles V.; Bray, George A.; and Swerdloff, Ronald S. "A Psychiatric Study of Patients Referred With a Diagnosis of Hypoglycemia." *American Journal of Psychiatry* 133 (1976): 290-94.

Frasier, S. Douglas; Hilburn, Jean M.; and Smith, Fred S. Jr. "Dwarfism and Mental Retardation: the Serum Growth Hormone Response to Hypoglycemia." *Journal of Pediatrics* 77 (1970): 136-38.

Fredericks, Edward J. and Lazor, Michael Z. "Recurrent Hypoglycemia Associated with Acute Alcoholism." *Annals of Internal Medicine* 59 (1963): 90-94.

Freinkel, Norbert; Arky, Ronald A.; Singer, David L.; Cohen, Alex K.; Bleicher, Sheldon J.; Anderson, John B.; Silbert, Cynthia K.; and Foster, Angela E. "Alcohol Hypoglycemia IV: Current Concepts of Its Pathogenesis." *Diabetes* 14 (1965): 350-61.

Gorman, C.K. "Hypoglycemia: A Brief Review." *Medical Clinics of North America* 49 (1965): 947.

Greenberg, Robert E. and Christiansen, Robert O. "The Critically Ill Child: Hypoglycemia." *Pediatrics* 46 (1970): 915-19.

Hafken, Louis; Leichter, Steven and Reich, Theodore. "Organic Brain Dysfunction as a Possible Consequence of Postgastrectomy Hypoglycemia." *American Journal of Psychiatry,* 132 (1975): 1321-24.

Hagler, Louis; Hofeldt, Fred D.; Lufkin, Edward G.; and Herman, Robert H. "Reactive Hypoglycemia." *Rocky Mountain Medical Journal* 70 (1973): 41-46.

Hamilton, Ronald D. "Hypercalcemia and Hypoglycemia in Adrenocortical Insufficiency." *Journal of Kentucky Medical Association* 74 (1976): 72-74.

Hamman, Louis and Hirschman, I.I. "Studies on Blood Sugar." *Archives of Internal Medicine* 20 (1917): 761-64.

Harper, Charles R. and Albers, William R. "Alcohol and General Aviation Accidents." *Aerospace Medicine* 35 (1973): 462-64.

Harper and Kidera, G.J. "Hypoglycemia in Airline Pilots." *Aerospace Medicine* 44 (1973): 769-71.

Harris, Seale. "The Diagnosis and Treatment of Hyperinsulinism." *Annals of Internal Medicine* 10 (1936): 514-33.

Hed, Ragnar and Nygren, Arne. "Alcohol-Induced Hypoglycemia in Chronic Alcoholics with Liver Disease." *Acta Medica Scandivica.* 183 (1968): 507-10.

Hofeldt, Fred D.; Dippe, Stephen; and Forsham, Peter H. "Diagnosis and Classification of Reactive Hypoglycemia Based on Hormonal Changes in Response to Oral and Intravenous Glucose Administration." *American Journal of Clinical Nutrition* 25 (1972): 1193-1200.

Hoffman, Richard H. and Abrahamson, Emanuel M. "Hyperinsulinism – a Factor in the Neuroses." *American*

Journal of Digestive Diseases 16 (1949): 242-47.

Ibarra, J.D. Jr. "Hypoglycemia." *Postgraduate Medicine* 51 (1972): 88-93.

Jay, Arthur N. "Hypoglycemia ... Symptoms, Diagnosis, Treatment." *American Journal of Nursing* 62 (1962): 77.

Jenner, B.M. "Alcohol-Induced Hypoglycemia." *Medical Journal of Australia* 2 (1979): 79-80.

John, Henry J. "Hyperinsulinism." *Medical Clinics of North America* 17 (1933-34): 979-85.

Jung, Y.; Corredor, D.G.; Hastillo, A.; Lain, R.F.; Patrick, D.; Turkeltaub, P.; and Danowski, T.S. "Reactive Hypoglycemia in Women, Results of a Health Survey." *Diabetes* 20 (1971): 428-34.

Kedes, Laurence H., and Field, James B. "Hypothermia, a Clue to Hypoglycemia." *New England Journal of Medicine* 271 (1964): 785-87.

Landmann, Heinz Richard and Sutherland, Richard L. "Incidence and Significance of Hypoglycemia in Unselected Admissions to a Psychosomatic Service." *American Journal of Digestive Disease* 17 (1950): 105-8.

Levine, Rachmiel. "Hypoglycemia." *Journal of American Medical Association* 230 (1974): 462-63.

Light, Marilyn Hamilton. "Hypoglycemia: an Overview." *Nursing Care,* February 1977.

Lister, John. "By the London Post." *New England Journal of Medicine* 297 (1927): 921-23.

Mackenzie, Thomas B. "Hypoglycemia and Alcoholism." *Psychosomatics* 20 (1979): 39-42.

Marks, Vincent. "Hypoglycemia and its Treatment." *Nursing Times,* August 10, 1972.

Marks. "The Measurement of Blood Glucose and the Definition of Hypoglycemia." *Hormone and Metabolic Research,* Supplement #7, Rome Symposium, 1976.

Marks. "Concluding Remarks: Hypoglycemia – Real and

Imaginary." *Hormone and Metabolic Research,* Supplement #7, Rome Symposium, 1976.

Marks and Medd, W.E. "Alcohol-induced Hypoglycemia." *British Journal of Psychiatry* 110 (1964): 228-32.

Martin, Lay and Hellmuth, George. "Hypoglycemia – a Study of 404 Patients Who Had No Insulin and Had This Common Finding." *American Journal of Digestive Diseases and Nutrition* 4 (1937): 579-87.

Mitchell, Allen A. "Relation of Body Weight and Insulin Dose to the Frequency of Hypoglycemia." *Journal of American Medical Association* 228 (1974): 192-94.

Mitchell, J. C. "The Posthumous Misfortune of Captain Bligh of the 'Bounty'." *Diabetes* 23 (1974): 919-20.

Moore, Henry; O'Farrell, W. R.; Malley, L.K.; and Moriarty, W.A. "Acute Spontaneous Hypoglycemia." *British Medical Journal* 2 (1931): 837-40.

Permutt, M. Alan. "Postprandial Hypoglycemia." *Diabetes* 25 (1976): 719-36.

Permutt; Kelly, John; Bernstein, Robert; Halpers, David H.; Siegel, Barry A.; and Kipnis, David M. "Alimentary Hypoglycemia in the Absence of Gastrointestinal Surgery." *New England Journal of Medicine* 288: 1206-10.

Portis, Sidney A. "Mechanism of Fatigue in Neuropsychiatric Patients." *Journal of American Medical Association* 12 (1943): 569-71.

Poulos, C.; Jean, D.; Stoddard, Donald; and Carron, Kay. "Hypoglycemia, the Hidden Menace in Alcoholism." *Nursing Care,* February 1977.

Raichle, Marcus E., and King, William H. "Functional Hypoglycemia: a Potential Cause of Unconsciousness in Flight." *Aerospace Medicine* 43 (1972): 76-78.

Rayner, Maurice, and Rogerson, C.H. "Paroxysmal Hyperinsulinism Due to Islet Adenoma of the Pancreas." *Lancet* 2 (1943): 476-79.

Ross, Harvey M. "Hypoglycemia." *Orthomolecular Psychiatry* 3 (1974): 240-45.

Runion, H. I. "Hypoglycemia, Fact or Fiction?" Paper read at International Conference Association for Children With Learning Disabilities, February 28-March 3, 1979, San Francisco.

Salans, Lester B., and Martin, Angela A. "What You Should Know About Hypoglycemia." *Pharmacy Times*, September 1978.

Salzer, Harry M. "Relative Hypoglycemia as a Cause of Neuropsychiatric Illness." *Journal of National Medical Association* 58 (1966): 12-17.

Seltzer, Holbrooke S. "Drug-Induced Hypoglycemia – a Review Based on 473 Cases." *Diabetes* 21 (1972): 955-61.

Seltzer. "Severe Drug-Induced Hypoglycemia: a Review." *Diabetes* 5 (1979): 21-28.

Service, F. John. "Hypoglycemia." *New York State Journal of Medicine* 2 (1978): 2122-23.

Shah, Sharfuddin. "The Sugar Connection." *Journal of the Kansas Medical Society* 494 (1976): 515-16.

Smelo, Leon S. "The Hypoglycemias, Their Diagnosis and Management." *Delaware Medical Journal* 45 (1973): 128-33.

Tintera, John W. "Office Rehabilitation of the Alcoholic." *New York State Journal of Medicine* 55 (1956): 3896-3902.

Tintera. "Stabilizing Homeostasis in the Recovered Alcoholic Through Endocrine Therapy: Evaluation of the Hypoglycemia Factor." *Journal of American Geriatrics Society* 14 (1966): 71-95.

Tintera. "The Hypoadrenocortical State and Its Management." *New York State Journal of Medicine* 56 (1955): 1869-76.

Trotter, Robert J. "Aggression: A Way of Life for the

Quolla." *Science News* 103 (1973): 76-77.

Varma, Surendra K. "Hypoglycemia in Infancy and Childhood." *Southern Medical Journal* 72: (1979) 57-64.

Ware, Stephen and Osborne, J.P. "Postoperative Hypoglycemia in Small Children." *British Medical Journal* 2 (1976): 499-501.

Wauchope, G. M. "Critical Review, Hypoglycemia." *Quarterly Journal of Medicine* 2 (1933): 117-33.

Whipple, Allen O. "Adenoma of Islet Cells with Hyperinsulinism." *Annals of Surgery* 101 (1935): 1299-1305.

Wilder, Russell M. "Hyperinsulinism." *International Clinics* 2 (1933): 1-18.

Wilkinson, Charles F. Jr. "Recurrent Migrainoid Headaches Associated With Spontaneous Hypoglycemia." *American Journal of the Medical Sciences* 218: 209-12.

Wolfe, Bernard M. and Powers, Rosemary. "Hypoglycemia." *Canadian Nurse,* October 1973, pp. 38-40.

Wright, John and Marks, Vincent. "Alcohol-induced Hypoglycemia." *Advnces in Experimental Medicine and Biology* 126 (1980): 479-83.

Young, Margaret. "Hypoglycemia, a Nursing Care Study." *Nursing Times* 66 (1970): 915-16.

About the Authors

Jinny Zack

The mother of three children, Jinny has spent the past several years conducting intensive research into the problem of hypoglycemia. In the process, she has made many appearances on television and radio talk shows and spoken to service organizations, informing listeners about the effects, diagnosis, and treatment of this national health menace. A member of Toastmasters International, she also has formed study clubs so that hypoglycemics can help each other.

Wilbur D. Currier, M.D.

More than thirty-five years ago, Dr. Currier discovered the importance of nutrition and became interested in metabolic disorders such as hypoglycemia. He has specialized in this field for more than twenty of these years. In 1971 he founded the International Academy of Metabology, a scientific group of doctors and other medical professionals, to study the problem.

Dr. Currier has taught at Harvard Medical School and is an associate professor emeritus of the University of Southern California. He has worked with thousands of hypoglycemics and has patients from all over the world.

This noted physician has published many scientific papers in the field of nutrition and metabolic preventive medicine. For twenty years, he has served as medical director of the Lancaster Foundation, a non-profit research corporation of California. Dr. Currier handled patients medically in research programs that resulted in two books by George Watson, Ph.D. (Harper & Row) concerning the chemistry of nervous and mental illness.

Dr. Currier has been a medical consultant to the U.S. Surgeon General, the government of India, and to Christian medical schools. A Fellow of the American College of Surgeons, he is a member of many medical societies, including the American Medical Association (AMA). He was for five years national secretary of the American Academy of Applied Nutrition.

Correspondence
to the authors may be
addressed to:
P.O. Box 5233
Carmel, CA 93921